MOLD MATTERS

Solutions and Prevention

By

Charles, Danielle, and Rachelle Dobbs,
Mold detection experts

Published by Dobbs Enterprises, Inc.
2945 Waumpi Trail
Maitland, FL 32751
Tel: 407-629-4820
www.MoldDetectionExperts.com

Published by Dobbs Enterprises, Inc.

First printing: August 2006

ISBN-13: 978-0-9717664-9-5
ISBN-10: 0-9717664-9-5

Printed in China

TABLE OF CONTENTS

iv

Foreword

Mold Matters is about education. It is about protecting health and building structures. If you suspect you have a mold problem in your home or building, don't panic. We are here to guide you through the entire process of ridding your home of mold. Our goal is to give you the knowledge you need to be in control of the situation. In the process you will save time and money and in the end things will be done correctly.

You will learn about the conditions required for mold growth, how to test air quality, and how to handle water intrusion. You will be guided on selection of experts with regards to inspection, air quality testing, and remediation.

If you don't have a mold problem, rejoice. This book will give you many tips on mold prevention. Knowing the conditions for mold growth is half the battle, implementing the steps necessary for prevention is the way to win the battle.

Whether you are a home or building owner, insurance adjuster, lawyer, apartment or hotel manager, or a member of a maintenance crew, this book is for you. The same principles that apply to homes, also apply to apartment complexes, retail spaces, and office buildings. This book can also be used as a teaching guide for novice mold inspectors who will gain practical knowledge with regard to mold assessment, its methodology and quality control.

We are certified mold inspectors - two of us have scientific backgrounds, MS physics and BA chemistry, and one is a former teacher and writer. Years of field experience observing, researching, and documenting water damage and mold conditions have led us to the conclusion that a large percentage of these cases could have been avoided by following simple preventive measures. The many questions from our clients and attendees at our lectures have motivated us to write this book to offer solutions to mold problems as well as giving specific mold prevention tips. Our goal is to empower you about mold matters.

ALL ABOUT MOLD

WHAT IS MOLD?

Mold matters are not simple and what you don't know can hurt you. All buildings are subject to developing mold at some point. Learning about mold will help you understand the conditions that lead to the problem and how to prevent it in the future. It will also help you understand why a three-step process in mold removal is necessary: 1) the mold assessment phase when the problem is being qualified and quantified, 2) the mold remediation phase when the problem is being fixed, and finally 3) the post-remediation phase which insures that the mold remediation has been carried out properly.

Mold, known scientifically as fungus, is a microscopic living organism and you might be surprised to learn it is not always bad. In fact, mold is highly beneficial to the environment. Its purpose in life is to help in the decomposition of dead organic material, and it has done so for millions of years. One of the earliest recorded references to mold remediation is found in the Bible, *Leviticus 14:33-53*, *"Cleansing From Mildew"*. At that time, priests had a major role in overseeing the treatment of a contaminated house. With the advent of science we have been able to analyze mold in order to understand its composition, structure, and habits, but research is still in its infancy with regards to the effects of mold on human health.

Mold spores are found everywhere – outdoors, inside homes and buildings, on our clothes, our hair, and on everything we touch. Fortunately, this is a normal occurrence and is simply part of living on earth. If mold disappeared tomorrow we would literally be living on a trash heap and all life on earth would become extinct in a matter of months.

Mold becomes a problem when it is growing inside homes or buildings. There are common molds found everywhere in nature, such as *Penicillium, Aspergillus, Cladosporium,* and *Basidiomycetes*, and there are molds that are rarely found outdoors, such as *Stachybotrys*. When common molds are found growing indoors in large quantities it can affect our health just as much as low levels of highly toxic mold. We must think of mold as being potentially toxic because mycotoxins have also been found in common mold. The levels of toxicity differ from one species to the next and the effect on human health varies according to the particular make-up of the individual.

Mold is highly resilient and has learned to adapt and thrive in many different climates. It can be found all over the globe, even in the harshest of environments, such as Antarctica. It has also been found in outer space. In 1988 it was discovered aboard the space station Mir and was later identified by Natalia Novikova, a microbiologist, from the Microbial Laboratory of the Russian government's Institute for Biomedical Problems. The fungi were from the genera *Aspergillus, Penicillium*, and *Cladosporium*.[1]

The scientific name for mold is fungi (singular: fungus). The classification is broken down further into genera (singular: genus). These are then classified further into species. All living things can be classified into seven kingdoms with fungi being number five:
- Protomonera
- Monera
- Protista
- Myxomycophyta
- Fungi
- Plantae
- Animalia

Fungus is both friend and foe. It fills a vital ecological niche, but when growing in a home or building it can compromise building materials and cause serious health problems.

In our everyday lives we benefit from its presence in our wine, our cheese, in our medicines, and in many commercial applications. Certain fungi provide us with penicillin and other antibiotics, and we eat other fungi, such as mushrooms and truffles. The yeast found in dough causes bread to rise and gives it a light texture, and the bubbles in champagne and beer give us something to cheer about.

Mold can also destroy precious artifacts from antiquity. Recently, forty varieties of mold have been identified eating away the 2,200-year-old army of terra cotta Chinese warriors found in the tomb of Qin Shihuang, the first emperor of China. A Belgian company has been commissioned to eradicate mold from 1,400 of the 8,000 life-sized statues of soldiers and horses.[2]

There are between 1.3 and 3 million species of mold. With that many, mold comes in all colors of the rainbow. Some molds can even change color based upon what they are feeding upon at the time.

DIFFERENT FACES OF MOLD

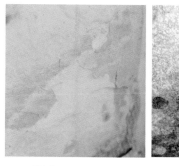 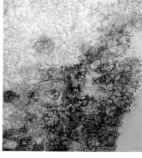 

Pink and orange   Grey and black   Green

 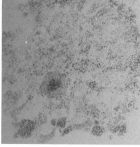 

Brown   Mosaic   Circular patches

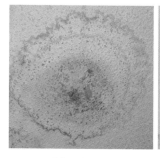  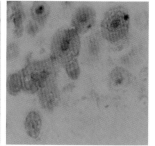

Like a rose   Lace   Galaxies

LIFE CYCLE OF MOLD

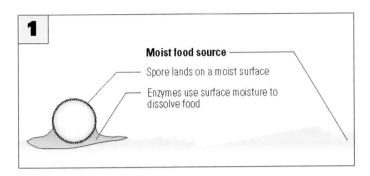

1

Moist food source

Spore lands on a moist surface

Enzymes use surface moisture to dissolve food

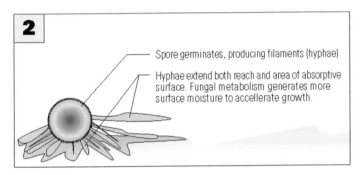

2

Spore germinates, producing filaments (hyphae)

Hyphae extend both reach and area of absorptive surface. Fungal metabolism generates more surface moisture to accellerate growth.

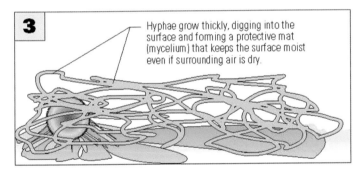

3

Hyphae grow thickly, digging into the surface and forming a protective mat (mycelium) that keeps the surface moist even if surrounding air is dry.

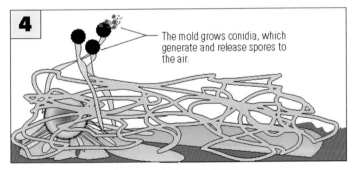

4

The mold grows conidia, which generate and release spores to the air.

Source: ASHRAE 2001
Humidity Control Design Guide[3]

To understand the physical structure of mold, a simple analogy with a dandelion can be made. Generally, a dandelion has a root structure, a stalk, and a seed releasing body at its top. Similarly, mold has a stalk-like structure with a spore-releasing body at the end. Mold even has root-like structures, although they act more as a digestive lattice than the benign roots of a dandelion. As wind blows across a dandelion, it releases its seeds to propagate itself; so too with mold. The least disturbance will cause the release of spores into the air. Some molds do not even need a breeze, and the spores simply fall around the main structure. This explains why mold often appears as small spots, or as a sort of ring. Other molds do not aerosolize easily and require active disturbance for spores to become airborne.

When spores floating in the air land in a suitable environment, they start to germinate, much like the seeds of a dandelion. The root-like structures of mold, called hyphae, then emerge. They anchor themselves into the substrate and start secreting enzymes to help it rot so that it can be absorbed. This cycle then repeats, ad infinitum.

Some people refer to mold as "mildew", but it is not the same thing. According to the Dictionary of Fungi (9th edition): mildew is a plant disease in which the pathogen is seen as a growth on the surface of the host. There are at least two kinds of plant pathogens, powdery mildews and downy mildews. These are common plant disease names given to distinct group of fungi. According to Dr. Payam Fallah of the Environmental Microbiology Laboratories in San Diego, "Mildews are considered plant pathogens and such terms should not be used in indoor air issues as they potentially confuse certain issues relating to microbial activities in indoor environment." As of this writing mold, with its spores and possible mycotoxins production, can directly affect human health, while mildew is a plant disease, which has no effect on human health.

WHAT ARE THE CONDITIONS FOR
MOLD GROWTH?

Spores need three things in order to grow: food, a surface to grow on, and water. When conditions are right, mold can start to grow and propagate in as little as 24 to 48 hours.

Of these three things, water is the only one we can control. For better or for worse, buildings will continue to be made from organic material: wood studs, pressboard, drywall, and many other common building materials that provide a food source for mold. Once water has been allowed to infiltrate into the home or building, time is the crucial element. The faster repairs are made, and the faster drying is implemented, the less likely that mold will gain a foothold.

The source of moisture can sometimes be difficult to locate, but there are a few common places to look. The air conditioning system is a common hiding place for mold colonies; check the air handler coils on a regular basis for any sign of mold growth. Take a look at the drain pan and condensate line to make sure they drain efficiently. The air conditioning system may not be functioning efficiently, causing the humidity in the building to be high enough for mold to grow. If this is the case, a supplemental stand-alone dehumidifier should be considered. Conversely, if the air conditioning is working too efficiently, a residence or building may be cool and moist, as a result of the system not running long enough to remove the humidity from the air. If a large home requires more than one air conditioning unit, care should be taken to ensure that the thermostats do not interfere with each other. If this happens, it can cause one air conditioning unit to switch off prematurely and mold may grow in one part of the house because of the elevated humidity. The humidity generated when you take a bath or shower makes it essential to turn on the bathroom exhaust fan or open a window; otherwise mold can start growing on the walls and ceiling. See the section on *MOLD*

PREVENTION - Tips on controlling moisture and water intrusion.

The relative humidity (RH) of the air is critical for mold. The total amount of moisture air can hold depends on the temperature of the air. Warmer air can hold a greater amount of moisture than colder air. The RH is based on two factors - the amount of moisture in the air and the temperature. The RH is defined as the ratio of the amount of water vapor in the air at a specific temperature to the maximum amount that the air can hold at that temperature, expressed as a percentage. At an RH of 100%, dew point is reached when moisture condenses on surfaces cooler than the surrounding air. The extent of fungal contamination is related to the indoor relative humidity. Below 30% relative humidity very little mold growth occurs, while at 70% conditions are optimum for mold growth. High humidity allows moisture to condense on cool surfaces, such as windows and sills. Moisture can also seep through walls, ceilings, basements, and concrete slabs.[4]

Pipe leaks are also notorious culprits leading to mold growth. Once a month, check under sinks, behind toilets, refrigerators, next to the water heater, etc. for any possible leaks. Copper pipes in older homes can develop pinhole leaks and with time these small leaks can cause tremendous damage. Be aware of the risk, and think about re-piping the house before problems occur. Grout and caulking around the showers and bathtubs should also be checked on a regular basis. Re-grouting is simpler than you think, and many hardware stores offer do-it-yourself classes. The worst kind of leak is an undiscovered one. See the section on MOLD PREVENTION.

Water intrusion from the outside is the next most likely source of moisture for mold. Missing roof shingles, cracks and breaks in walls, leaky windows, sprinklers less than two feet away from the outside wall, blocked gutters, vines, and poor drainage are all common culprits for water intrusion. Walk around the outside of your home, and check your attic at least three times a year to keep an eye out for any of these potential problems. The fixes are simple enough; a leaky roof warrants an inspection from a consultant. Likewise, get leaky doors and windows repaired. Sprinklers too close to the house should be moved away from the building or sprinkler splash guards can be purchased at most hardware stores. Gutters are often forgotten, causing rotting debris and, eventually, leaks into the building. Although the look is beautiful, decorative vines on the outside wall will cause façade deterioration and fine cracks will give water a conduit to the inside.

Seasonal allergies caused by mold can be a sign of a larger problem, such as mold growing inside the air conditioning duct system. When weather forces the change from air conditioning to heat, the dry, warm air coursing through ductwork can dry out condensation, and therefore mold. When the source of moisture stops, some species will release more spores into the air. If your allergies hit a peak around that time of year, it could mean mold is growing in the ductwork.

HOW MOLD AFFECTS HEALTH

Mold can be classified into three broad types as far as health effects are concerned. The first category is allergenic molds, which cause allergic or asthmatic reactions, but do not usually cause permanent health effects in most healthy, active people. There are pathogenic molds, which can cause serious health problems in those who are more susceptible. And finally, there are toxic molds that can cause serious health problems in everybody. The severity of these problems differs depending on age, immune system, and sensitivity. Children, the elderly, and people with depressed immune systems due to cancer, organ transplants, or AIDS, can become very sick when exposed to higher than normal levels of mold. Even some healthy individuals happen to be very sensitive to mold and are unable to tolerate a slight elevation of mold spores.

Most reactions to mold are due to inhaling spores that are floating in the air. Dr. Burge warns that nonviable (dead) spores retain their allergenic properties.[5]

The health effects of mold are varied, from mold growing *on* you, such as simple athlete's foot, to far more serious infections, such as *Aspergillosis*, which is caused by Aspergillus mold growing inside the body. *Aspergillosis* is a non-contagious disease of the genus *Aspergillus* that affects humans and pets equally, particularly in immuno-compromised hosts. The infection starts with fungal inhalation and is then dispersed to tissues and organs. The disease can affect the eyes, nose, heart, lungs, intestines, kidneys, and more.

In recent years, so called "mold dogs" have been in the news as celebrity and novelty. Their job is to sniff and locate mold, and thus they are being exposed to mold on a daily basis. What people may not know is that long nose dogs are at relatively high risk of developing *Aspergillus sinusitis* or *nasal Aspergillosis.*[6] We personally have heard of two mold remediators who had to put their dogs to sleep after two years because both developed nose cancer.

Contaminated food can also affect animals. As recently as December 31, 2005, the Associated Press reported that a prominent pet food company had to recall 19 varieties of dog and cat food because some of the food had been contaminated by *Aflatoxin*, a chemical produced by mold. Twenty three animal deaths have been linked to the contaminated pet food.

A 1994 Harvard study of 10,000 homes found that half had "conditions of water damage and mold associated with a 50 to 100% increase in respiratory symptoms".[7]

The condition called St. Anthony's Fire, which killed thousands of people during the Middle Ages and in the ancient world, is caused by *ergotism*, or the eating of bread made with grain contaminated with the *ergot fungus*. St. Anthony's Fire is a gruesome disease characterized by rotting flesh, hallucinations, convulsions, and dry gangrene. People once believed that by making supplications and pilgrimages to St. Anthony and his shrine, a cure could be granted. Interestingly enough, these actions often worked because the pilgrims, traveling from place to place, would cease ingestion of the contaminated bread.[8]

The fungus *Phytophthora infestans* caused the Irish potato famine of the 1840s. The famine caused a million people to die from disease or starvation. The impact of mold on crops is staggering; it is said that over one third of all worldwide crop losses are caused by fungal disease.[9]

During the Second World War, as well as for some years after, a condition named *Alimentary Toxic Aleukia* struck a large percentage of the Siberian populace. The hemorrhaging, low leukocyte count, and high fatality rate were caused by mold. The war had caused a manpower shortage, which in turn resulted in the grain being harvested in the spring instead of the fall. The grain had moldered in the fields and had become extremely toxic when eaten.[9]

An unusually high incidence of esophageal cancer among the Xhosa people in Africa seems to be linked to the native beer of their diet. This drink is commonly made with grain contaminated with the fungus *Fusarium moniliforme*.[9]

Even in our modern day, mold still causes problems in both livestock and crops. In 1960 the death of more than 100,000 turkeys in England, and the subsequent investigation, brought about the discovery of *Aflatoxin* on peanuts used as animal feed. *Aflatoxin*, caused by the common mold *Aspergillus flavus*, was found to have alarmingly potent carcinogenic properties and, as a result, limits of contamination were immediately implemented. Mold continues to harm livestock, causing "Hole in the Head" disease in horses, facial eczema and lupinosis in sheep, and "Grass Staggers" in cattle to name just a few.[9]

The Mayo Clinic, a renowned research institution has pioneered several studies on chronic sinusitis to determine whether mold spore exposure and inhalation played a part in the disease. A research project conducted in 1999 indicated a link between chronic sinusitis infections and fungus (mold) in 93% of the subjects.[10]

Dr. Harriet Burge from The Environmental Microbiology Laboratory has found that different fungi have different temperature requirements for optimal growth. She states:

> In tropical and subtropical places where both heat and moisture are present . . . incidence of fungal infections (including sinus infections) tends to be higher in these areas in part because the fungi that can withstand human body temperatures are more abundant than in temperate climates.

According to a recent survey by the National Center for Health Statistics 14.1% of the U.S. population suffers from chronic rhinosinusitis. This means that 1 in 7 people suffer from the disease.[11]

In 2005 researchers from the Mayo Clinic found that fungi plays a large role in chronic rhinosinusitis. In fact, the findings indicates that chronic rhinosinusitis is a result of a fungal driven inflammation rather than a bacterial infection.[12]

Indoor air quality problems in schools affect both students and teachers. The following statistics were published on February 2, 2005 by the IEQ Review:
- One in five schools in America has indoor air quality problems.
- Asthma accounts for 14 million missed school days each year.
- The rate of asthma in young children has risen by 160 percent in the past 15 years.
- 1 out of every 13 school-age children has asthma.[13]

The Center for Indoor Environments and Health at the University of Connecticut states "the most common types of illnesses directly related to mold are type I responses of allergic rhinitis and asthma." They go on to say "… allergic inflammation can trigger bronchospasm, chest tightness, and shortness of breath, leading to either new onset of asthma or asthma exacerbation in sensitized individuals."[14]

Poor maintenance in schools and lack of money are often cited as excuses for mold problems, but little is done about it. This does not only pertain to public schools; some private schools are just as bad. Many university dorms, regardless of school prestige, are in poor condition and some harbor mold. Students accept these conditions as status quo and fail to complain. This situation does not have to be. If money can be found to modernize a gym and re-sod the school lawn, money can be found to maintain buildings properly. Air quality should be a priority of any institution, and parents and teachers should demand it.

Many apartment buildings have mold problems and maintenance personnel know little or nothing about mold. Their lack of knowledge can sometimes cause them to wipe and paint over a contaminated area in hopes that the problem will go away. Predictably within a few months mold will reappear. This is because the hyphae (the root system) are still embedded in the wall. Unless the source of moisture is corrected and the contaminated area physically removed, mold will continue to grow.

Sometimes, people are not aware that a mold problem exists in their home or workplace, and when they develop allergy-like symptoms they seek the help of an allergist. If medical tests are negative or inconclusive, allergy specialists should recommend having the home or workplace tested for mold. If the levels of mold spores are elevated, the problem can be found and fixed, and with luck the person will regain his or her health without suffering permanent damage.

With regard to toxic mold, the United States Environmental Protection Agency (EPA) states:

> Molds can produce toxic substances called mycotoxins. Some mycotoxins cling to the surface of mold spores; others may be found within spores. More than 200 mycotoxins have been identified from common molds, and many more remain to be identified. Some of the molds that are known to produce mycotoxins are commonly found in moisture-damaged buildings. Exposure pathways for mycotoxins can include inhalation, ingestion, or skin contact. Although some mycotoxins are well known to affect humans and have been shown to be responsible for human health effects, for many mycotoxins, little information is available.

> *Aflatoxin B1* is perhaps the most well known and studied mycotoxin. It can be produced by the molds *Aspergillus flavus* and *Aspergillus parasiticus* and is

one of the most potent carcinogens known. Ingestion of *aflatoxin B1* can cause liver cancer. There is also some evidence that inhalation of aflatoxin B1 can cause lung cancer. *Aflatoxin B1* has been found on contaminated grains, peanuts, and other human and animal foodstuffs. However, *Aspergillus flavus* and *Aspergillus parasiticus* are not commonly found on building materials or in indoor environments.

Much of the information on the human health effects of inhalation exposure to mycotoxins comes from studies done in the workplace and some case studies or case reports. Many symptoms and human health effects attributed to inhalation of mycotoxins have been reported including: mucous membrane irritation, skin rash, nausea, immune system suppression, acute or chronic liver damage, acute or chronic central nervous system damage, endocrine effects, and cancer. More studies are needed to get a clear picture of the health effects related to most mycotoxins. However, it is clearly prudent to avoid exposure to molds and mycotoxins.

Some molds can produce several toxins, and some molds produce mycotoxins only under certain environmental conditions. The presence of mold in a building does not necessarily mean that mycotoxins are present or that they are present in large quantities.

The Internet has a wealth of information on the subject of mold. One can find cases in which mold has caused irreversible and permanent damage in humans and pets. If you have concerns or questions relating to your health, do not try to diagnose your health problems with information found on the Internet or information contained in this book - see your doctor. If you want to ascertain whether a mold problem exists in your home, get it tested by a professional mold inspector.

HOW MOLD AFFECTS BUILDING STRUCTURES

We should not be surprised to learn that mold destroys building structures since its purpose in life is to break down organic material. Yet, we tend to build with organic materials. With time we learn from Mother Nature and we improve building codes when certain building practices prove to be contrary to the laws of nature.

The increase in incidence of mold contamination in recent times can be attributed in great part to energy conservation measures. This has made our homes much tighter than they used to be. In so doing, natural ventilation has been cut down, which would otherwise help dry water infiltration, condensation, or leaks when they happen. Other factors contributing to mold are cheaper building materials, poor workmanship, leaving building materials on job sites unprotected from rain, and cutting down on time allowed to cure materials. All this and more has contributed to making homes and buildings more susceptible to mold.

Wood rot

When mold attacks solid pieces of wood, it takes longer to deteriorate, because its cells are not fractured. Pressboard, on the other hand, has fractured cells and cellulose-based glues (sugar). This means that the rate of decomposition in pressboard is much higher than in solid wood and the glues used are candy for mold.

Impermeable materials, such as vinyl wallpaper can also lead to mold problems. Condensation can form on the back of the vinyl because of temperature differences between the inside and outside air. Where there is water and food (condensation and wallpaper glue), mold can settle in.

There are two areas in a home especially prone to mold – bathrooms and closets. The high humidity caused by daily baths and showers, coupled with skin cells and body oils (mold food) makes a perfect environment for mold to grow. Mold will settle in and deteriorate grout, caulking, paint, and sheetrock. The high humidity in closets is usually due to poor ventilation, or being too cold.

Any plumbing system has a potential for leaks. Think for a moment what we do to shower and bathtub plumbing – we enclose the pipes inside walls and forget about them. This means that if a leak occurs it will not be detected immediately. Over time building materials will deteriorate and mold will grow. Homeowners having closets backing up to shower stalls should consider themselves lucky rather than having the plumbing backing up to the outside wall. Neither condition is ideal, but the former is somewhat better than the latter. Leaks are more easily detected and repaired through a closet wall. On the other hand, if the plumbing backs up to an outside wall, there is no way to fix anything without tearing down the entire shower. Architects should never design a bathtub or shower backing up to an outside wall. A mean of accessing the plumbing should always be in the design plan. See *MOLD PREVENTION*.

There are some exterior finish systems that do not allow water to drain, such as EIFS, or Exterior Insulating Finishing Systems. This can be bad news. In an ideal, perfect, and controlled environment, EIFS works great. But, our world is not perfect, nor controlled. Water comes through the faux stucco and has nowhere to go. The result is wood rot and mold. To check whether an exterior finish is EIFS, do a "tap" test. Tap on the outside of the exterior wall, and if it sounds hollow, you may have EIFS. Some new exterior finishing systems are being improved to allow for drainage.

There are three main strategies to minimize the risk of moisture damage, says Dr. Lstiburek:
1. Control of moisture entry
2. Control of moisture accumulation
3. Removal of moisture.

This is easier said than done. Dr. Lstiburek summarizes the problem as follow:

> Strategies effective in the control of moisture entry, however, are often not effective if building assemblies start out wet, and in fact can be detrimental. If a technique is effective at preventing moisture from entering an assembly, it is also likely to be effective at preventing moisture from leaving an assembly. Conversely, a technique effective at removing moisture may also allow moisture to enter. Balance between entry and removal is the key.[15]

There are two sides to every coin. Searching for the perfect building construction methods must be left to experts, but real estate owners must assume the responsibility of building maintenance, and must also practice prevention. It is a fact of life that things deteriorate over time. Proper maintenance and timely repairs are the answers. See *MOLD PREVENTION*.

Water can do an enormous amount of damage to a building. Wind-driven rain during hurricanes seems to defy gravity; water can enter homes sideways, up under the soffits, and through cracks you didn't even know you had. Homeowners should prepare their homes for the inevitable and prevent water entry by sealing the walls, cracks, and other openings in the building.

IS THERE ANYTHING THAT CAN KILL MOLD?

The answer is: yes and no! There are non-commercial grade products, such as germicidal bleach, that are effective in killing some fungi and can be used as a cleaning solution. But, it should only be used on non-porous surfaces, such as metal or glass, and not on materials such as sheetrock/drywall, which is porous. Do not attempt to remediate large mold contaminated areas by yourself. You could affect your health and contaminate your entire home – leave it to professional mold remediators. They have the know-how, the right equipment and protective gear. As a rule, the mold you see on the wall is the tip of the iceberg and the infestation is often coming from within the wall cavity. If this is the case the wall has to be remediated – meaning part of the sheetrock, or the entire wall must be removed. Contaminated wood should be replaced whenever possible, or be soda or dry ice blasted, or sanded to remove any embedded hyphae.

Professional mold remediators have chemicals (fungicides) they use as final cleaning solutions during remediation. Mold remediation is the physical removal of the mold-infested medium performed under safe conditions. Do not confuse disinfectants with fungicidal products. Disinfectant kills (some) germs, fungicide kills (some) fungus. Both should be used with all necessary precautions.

There are many claims about all kinds of devices that are supposed to kill or prevent mold, such as "ozone generators". The EPA warns:

Manufacturers and vendors of ozone devices often use misleading terms to describe ozone. Terms such as "energized oxygen" or "pure air" suggest that ozone is a healthy kind of oxygen. Ozone is a toxic gas with vastly different chemical and toxicological properties from oxygen. Several federal agencies have established health standards or recommendations to limit human exposure to ozone.

WHY CAN'T I JUST GET AN AIR CLEANER?

We have all seen TV commercials for indoor air cleaners that promise to "get rid of mold!" (whether you have it or not). Know one thing - nothing will "get rid of mold", except the physical removal thereof.

Having said that, we must tell you that air cleaners equipped with HEPA filters work well in removing excess mold spores from the ambient air of a room. HEPA stands for High Efficiency Particulate Air, which traps airborne particles as small as 0.3 microns.

With respect to mold, we recommend stand-alone HEPA filter air cleaners only as a temporary fix and only in two instances. They are helpful for homeowners who know they have mold, but must wait until professional mold remediators can come to remediate the situation. While they wait, a HEPA filter unit will help clean the air of mold spores and will therefore help protect their health. Another instance is when tenants are unable to leave a mold-infested apartment, perhaps because of a landlord unwilling to help, or they are financially unable to move out. An air cleaner will aid in protecting the health of the tenants until they can move to a new place.

An air cleaner should not be used as a band-aid for mold issues. It is always preferable to fix the problem rather than mask it. Once the problem is repaired, the amount of mold spores inside a home will be less than or equal to the levels found outside. So, there would not be any reason to obtain an air cleaner to remove mold spores from the air.

If you are allergic to pollen and live in an area with excessive vegetation, a HEPA filter unit may help your allergies while you are inside your home.

HEPA filter air cleaners come in all kinds of shapes and sizes. Usually the unit is a freestanding electrical appliance that can be plugged in anywhere. They can be purchased at major hardware stores and some department stores carry them as well.

Television commercials are scaring people into believing that everybody has mold problems. This is not true. Many air quality tests we have performed showed normal levels of mold spores. We have seen instances where people were so paranoid about mold that they had purchased a HEPA filter unit for each room of their house. We have tested their homes after all filtering devices were off for 48 hours to get a true reading and found the air quality to be normal. One lady felt she had to leave her home and stay with a friend during these 48 hours because she felt that she would not be able to breathe without her HEPA filters. Test results showed that the readings in her house were ideal without the units, and the levels of mold spores were less than the levels found outside for all types of mold.

If you suspect you have a mold issue get the air tested to ascertain whether a problem exists. If laboratory results indicate a mold problem, get a mold inspection to locate the source of water and mold infestation, and get the problem fixed. If the air quality is normal, don't waste your money with a HEPA filter unless you buy it for a specific purpose other than a permanent, so-called, solution for mold. See *MOLD PREVENTION*.

HOW TO HANDLE WATER INTRUSION

Hurricanes

Floods

Groundwater infiltration

Sewage backup

WHAT TO DO WITH WET
BUILDING MATERIALS AND FURNISHINGS

In previous chapters we have learned that the 24 to 48 hours following water intrusion is critical. If building materials are dried during that time, chances are good that mold will not develop. This outcome is possible when a small leak is involved and when it is detected immediately.

When environmental disasters happen, such as hurricanes, floods, groundwater infiltration, sewage backup, or other causes, the amount of water entering a home can be overwhelming. You must take action during the first 24 hours. If drying companies cannot get to you because of a backlog, don't wait. There are things you can do to prevent any further damage to the building.

With regards to your possessions, certain things must be removed from the building and some possibly discarded, such as carpet, couch, shoes, etc. However, it is important to save the discarded items as proof until your insurance adjuster has had a chance to verify your claim.

The Federal Emergency Management Agency (FEMA) has much information on its website about many hazardous situations, including floods, hurricanes, tsunamis and more. Get acquainted with their services now: www.fema.gov/hazards/floods/whatshouldidoafter.shtm.

With regards to flooding, FEMA offers the following tips:

- FIRST STEP: If your home has suffered damage, call the agent who handles your flood insurance to file a claim. If you are unable to stay in your home, make sure to say where you can be reached.

- To make filing your claim easier, take photos of any water in the house and save damaged personal property. If necessary, place these items outside the home. An insurance adjuster will need to see what's been damaged in order to process your claim.

- Check for structural damage before re-entering your home. Don't go in if there is a chance of the building collapsing.

- Upon re-entering your property, do not use matches, cigarette lighters or other open flames since gas may be trapped inside. If you smell gas or hear hissing, open a window, leave quickly, and call the gas company from a neighbor's home.

- Keep power off until an electrician has inspected your system for safety.

- Check for sewage and water line damage. If you suspect damage, avoid using the toilets and the taps and call a plumber.

- Throw away any food -- including canned goods -- that has come in contact with floodwaters.

- Until local authorities declare your water supply to be safe, boil water for drinking and food preparation.

The Institute for Environmental Assessment (IEA) from Minneapolis, Minnesota and the Department of Environmental Health & Safety (DEHS) at the University of Minnesota have designed the following protocol for handling wet building materials and furnishings:

1.0 GENERAL

Inventory all water-damaged areas, building materials, and furnishings. Special attention should be given to identify carpet under cabinets, furnishings, etc. Also, utilize a moisture meter to identify the extent of water damage to drywall.

Once the water-damaged inventory is completed, document the type of water damage, i.e. clean water (potable sources), steam, unsanitary water (rain, ground water), or contaminated water (sewage).

NOTE: If a steam leak is a cause of moisture damage, additional consideration for the chemicals added to the steam is needed. Even though steam is relatively clean, steam treatment chemicals may remain in the materials affected by the steam leak after the moisture has dried.

2.0 ELECTRICAL

Consider all wet wiring, light fixtures, and electrical outlets to be shock hazards until they have been checked by a building inspector and/or electrician. Until then, turn the power off in the area of water damage. [Note: Only persons knowledgeable about electrical shock hazards should shut the power off.] Replace all electrical circuit breakers, GFIs (Ground Fault Interrupters), and fuses that have become wet. Switches and outlets that are wet can be cleaned and reused but, when in doubt, replace them.

All electrical motors, light fixtures, etc. that were wet need to be opened, cleaned, and air dried by a qualified person. Before being put back into service, inspect the motors, light fixtures, etc. to ensure there is no visible moisture/water droplets.

3.0 CEILING TILE

Remove and dispose of all wet ceiling tiles within 24-48 hours of water damage. The only exception would be if ceiling tile has become wet due to a small steam leak and the shape of the tile has not been altered. In this situation, the ceiling tile can be air dried and reused. In situations where the tile has been impacted by unsanitary water (>24 hours or previous water damage) or contaminated water, controlled methods should be utilized for removal and disposal. Controlled methods can range from personal protection to full abatement under negative air conditions dependent on the extent of the damage.

4.0 DRYWALL/LATHE PLASTER

Remove and replace all water-damaged drywall and insulation within 24 hours. If the drywall is not removed within 24 hours, if previous water damage has caused microbial growth, or if the sheetrock has been damaged by unsanitary or contaminated water, then extensive controls will be necessary for the removal process. Use a moisture meter and cut sheetrock at 12" to 48" above the moisture mark.

All hard surfaces such as block walls, etc. should be scrubbed with a mild detergent followed by a rinse of the surface using a solution of 1/4 cup bleach per gallon of water. Follow this with a clean water rinse. After work is completed, turn the heat **UP** (if possible) and utilize dehumidifiers to dry the area.

BLEACH CAUTION: The chlorine in the bleach may cause corrosion, therefore, avoid using on metal surfaces. Instead, use the aforementioned cleaning procedure with only a wash with a mild detergent. Also, bleach may fade colors. Therefore, test the bleach solution in an inconspicuous location before proceeding. **USE BLEACH IN A WELL VENTILATED AREA. DO NOT MIX BLEACH WITH OTHER CLEANING CHEMICALS, ESPECIALLY THOSE CONTAINING AMMONIA. POISONOUS VAPORS WILL RESULT.**

Wet lathe and plaster will leach the minerals from the wall and form a chalky surface. The loose material on the surface will need to be removed under controlled conditions and the surface allowed to dry. The surface can then be painted with an antimicrobial paint.

If the plaster/lathe wall develops a strong odor, with or without visible mold growth, remove people from this area of the building. Eliminate the source of the water and replace the water-damaged plaster. During replacement of the plaster/lathe, the following general procedures are recommended.

Controls to limit the spread of contamination include: setup of critical barriers, create a negative air differential, and use appropriate respiratory protection, gloves, and overalls for the workers. Excellent personal hygiene, including hand washing and showering after work in the area, is also recommended.

5.0 FURNITURE

Dispose of upholstered furniture that has become wet due to floods, roof leaks, sewage backup and groundwater infiltration. Upholstered furniture damaged by steam* leaks (See NOTE in #1.0 above) or direct contact with potable water should be dried within 24 hours and monitored for fungal growth and odors.

Hardwood furniture or laminate furniture whose laminate is intact should be air dried and cleaned with a detergent solution and rinsed with clear water and dried. See BLEACH CAUTION in #4 above.

Laminate furniture experiencing delamination should be disposed of because the pressed wood under the laminate absorbs water readily and is hard to dry.

Furniture made of particle board or pressed wafer board should be discarded. The exception would be if the furniture has become wet due to a steam leak or direct contact with potable water. In this situation, the furniture can be dried and monitored closely for fungal growth/odor. If fungal growth occurs or the furniture develops an odor, the particle board/pressed wood furniture should be discarded.

6.0 FILES/PAPERS

Remove and dispose of nonessential wet files and paperwork. The exception again would be if the moisture was due to steam leaks (See NOTE in #1.0 above); then these can be dried. Essential wet paper from water-damaged areas should be moved to a location where it can be dried, photocopied and then discarded. Professional conservators should be contacted for information on handling these types of wet products: American Institute of Conservation, 202-452-9545, Fax: 202-452-9328.

If a large amount of files and paperwork cannot be dried within 24-48 hours, essential files/paperwork may be rinsed with clean water and temporarily frozen until proper drying can be completed. Discard any paper products that develop mold.

7.0 CARPET

Any carpet that has been contaminated over a large area with sewage backup should be discarded under controlled negative air conditions and the entire area disinfected with bleach and water (or hospital-grade detergent).

Small areas of carpet contaminated with sewage backup may be cleaned using the procedure listed for other sources of water.

Carpet that has become wet from floods, roof leaks, steam leaks, potable water leaks, and groundwater can be treated as per the following:

Carpet wet less than 24 hours
- Remove all materials (e.g. furniture, file cabinets) from the carpet.

- Extract as much water as possible from the carpet using wet vacuums.

- Shampoo the carpet with a dilute surfactant.

- Soak with a 1/4 cup bleach to 1 gallon water solution.

Maximum concentration: a solution of 1 part bleach to 10 parts water. See BLEACH CAUTION in #4 above.

- It is preferable not to use a biocide. If a decision is made to use a biocide, consult a microbiologist. Reason: People may have a reaction to biocides. Often, quaternary amine compounds will be used as a biocide/cleaning compound. The compound may reduce levels of bacteria, but is often ineffective in killing fungal spores.

- Rinse and extract the carpet with clean water to remove detergent/bleach residues.

- Commercial steam cleaning of carpet can be used in place of bleach. The vacuum system is housed in a truck. The water is heated near the boiling point and is used to clean the carpet. Dry the carpet within 12-24 hours of treatment. After work is completed, increase the room temperature and use commercial humidifiers, floor fans, or exhaust fans to aid in drying the carpet.

Carpet wet more than 48 hours
- Wintertime:
  If carpet becomes wet during the winter with relative clean water, the previous protocol can be used to manage the carpet and salvage it.

- Summertime:
  Drying the carpet is usually more difficult in the summertime than the winter if the carpet is not in an air-conditioned space or dehumidifiers are not available. Water-damaged carpets in humid environments often do not dry adequately. Disposal of water-damaged carpets in humid environments is often the best option.

8.0 TESTING

Air and building material/furniture testing for microorganisms may be performed immediately after the water problem and periodically thereafter by a trained environmental health professional to ensure that there is no excessive human exposure to microbial growth. Post cleanup clearance sampling and inspection are necessary to ensure no excessive concentration of microbes will exist in the building.

Areas that are harder to access for cleaning should be specifically tested and any areas of carpet that had material on top should be tested.

If the carpet develops an odor or visible mold growth is apparent, the carpet should be removed under controlled conditions. If mold-sensitive persons react when entering a space with previously water-damaged carpet, with no odor or visible mold growth, the carpet should be tested or discarded under controlled conditions.

An exception to the testing would be materials that had moisture infiltration for the first time and are being discarded. However, the decision on testing needs to be made on a case-by-case situation after all the variables have been considered.

Note on Personal Protective Equipment

If testing has confirmed microbial growth on previously wet materials, then appropriately trained personnel with appropriate respiratory and personal protection should be used. Negative air enclosures may also be set up for limiting cross-contamination from damaged areas.

The following six flowcharts recap pictorially a decision process regarding water-damaged building materials.

Ceiling Tile

Categorize the type of water damage.

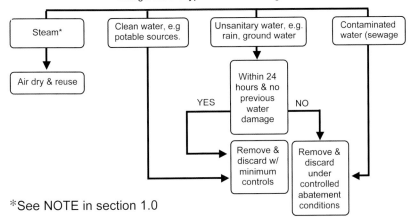

| Steam* | Clean water, e.g potable sources. | Unsanitary water, e.g. rain, ground water | Contaminated water (sewage |

Steam* → Air dry & reuse

Unsanitary water → Within 24 hours & no previous water damage

YES → Remove & discard w/ minimum controls

NO → Remove & discard under controlled abatement conditions

*See NOTE in section 1.0

Sheetrock

Categorize the type of water damage.

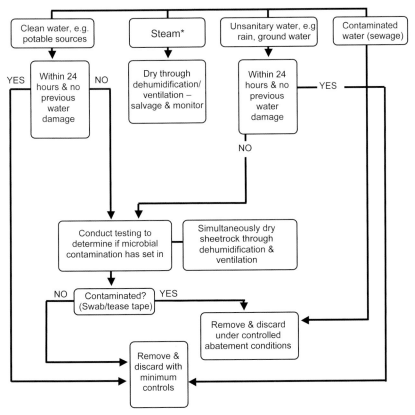

| Clean water, e.g. potable sources | Steam* | Unsanitary water, e.g rain, ground water | Contaminated water (sewage) |

Clean water → Within 24 hours & no previous water damage (YES / NO)

Steam* → Dry through dehumidification/ ventilation – salvage & monitor

Unsanitary water → Within 24 hours & no previous water damage (YES / NO)

Conduct testing to determine if microbial contamination has set in

Simultaneously dry sheetrock through dehumidification & ventilation

Contaminated? (Swab/tease tape) — NO / YES

Remove & discard under controlled abatement conditions

Remove & discard with minimum controls

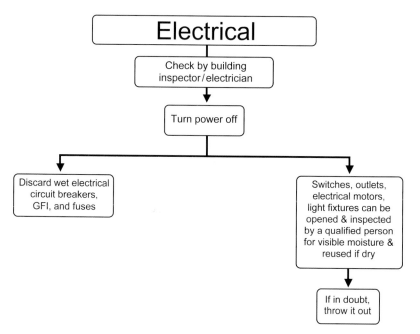

Electrical

Check by building inspector / electrician

↓

Turn power off

Discard wet electrical circuit breakers, GFI, and fuses

Switches, outlets, electrical motors, light fixtures can be opened & inspected by a qualified person for visible moisture & reused if dry

↓

If in doubt, throw it out

*See NOTE in section 1.0

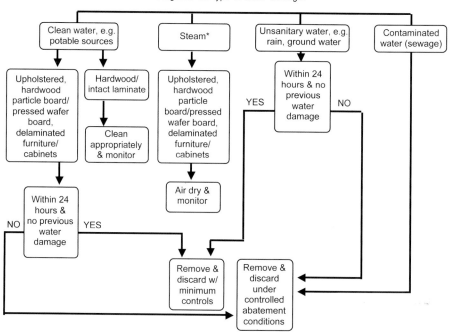

Furniture/Casework

Categorize the type of water damage.

Clean water, e.g. potable sources

Steam*

Unsanitary water, e.g. rain, ground water

Contaminated water (sewage)

Upholstered, hardwood particle board/ pressed wafer board, delaminated furniture/ cabinets

Hardwood/ intact laminate

Upholstered, hardwood particle board/pressed wafer board, delaminated furniture/ cabinets

Within 24 hours & no previous water damage

Clean appropriately & monitor

Air dry & monitor

Within 24 hours & no previous water damage

YES — NO

NO — YES

Remove & discard w/ minimum controls

Remove & discard under controlled abatement conditions

36

Paper/Files

Categorize the type of water damage.

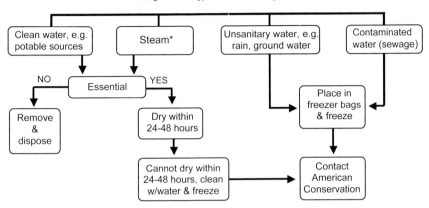

Carpet

Categorize the type of water damage.

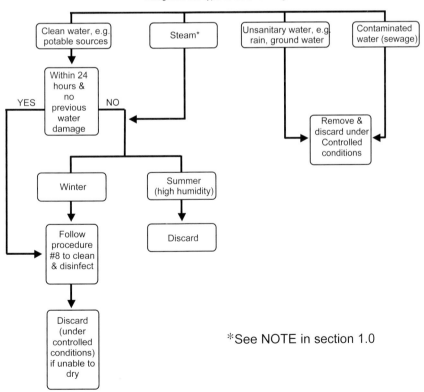

*See NOTE in section 1.0

ASSESSING INDOOR AIR QUALITY

WHAT ARE INDOOR POLLUTANTS?

Indoor air pollution involves a great deal more than mold. There are many indoor pollutants that can produce health effects similar to mold. The difference lies in people's acceptance of conditions that are less than desirable.

Occasionally, we are called to collect mold samples in a home or building because of odors or discomfort felt by the occupants. Once the laboratory results come back showing normal levels of mold spores, we must turn our attention to other concerns.

There are two main categories of indoor contaminants:

- Biological contaminants: Mold, bacteria, pollen, and viruses. Certain molds produce microbial volatile compounds (MVOC) and some of these are known to be toxic. Mold gives off a musty odor, while bacteria gives off an odor similar to rotten eggs.

- Chemical contaminants: Adhesives, carpeting, upholstery, manufactured wood products, copy machines, pesticides, and cleaning agents may emit volatile organic compounds (VOCs), including formaldehyde. The smell varies according to the chemicals. Tobacco smoke can also contribute to high levels of VOCs.

The Environmental Protection Agency states that certain VOC's are carcinogens. It warns that low to moderate levels of multiple VOCs may produce acute reactions and some can cause chronic and acute health effects at high concentrations. See *IMPROVING AIR QUALITY WITH PLANTS.*

"Sick Building Syndrome", or SBS, refers to certain symptoms that affect some occupants while they are in a building and disappear when they leave. The symptoms cannot be traced to specific pollutants or sources. It is believed that energy conservation measures following the energy crunch of the 70s have contributed to SBS. The "superinsulation" improved energy efficiency while reducing fresh air exchange. In the late 1970s, workers began complaining of various health problems, such as itchy eyes, skin rashes, drowsiness, respiratory and sinus congestion, headaches, and other allergy-related symptoms. Dr. Wolverton offered the following explanation:

> The airtight sealing of buildings contributed significantly to the workers' health problems. Similarly, synthetic building materials, which are known to emit or "off-gas" various organic compounds, have been linked to numerous health complaints.[16]

In 1989, the EPA submitted a report to Congress in which more than 900 organic chemicals were identified in newly constructed buildings. The report warned that some chemicals were in amounts one hundred times the norm. Today, the EPA states that "… contaminant concentration levels rarely exceed existing standards and guidelines even when occupants continue to report health complaints."[17] Considering the large number of VOCs found in buildings, collecting air samples of volatile compounds may or may not identify possible causes of occupants' discomfort.

While working with the National Aeronautics and Space Administration, Dr. Wolverton conducted many experiments involving the use of indoor plants to help remove indoor contaminants. His research concluded that specific plants play a role in removing specific volatile compounds, such as formaldehyde, benzene, trichloroethylene (TCE), and others. See *IMPROVING AIR QUALITY WITH PLANTS.*

HOW TO TEST INDOOR AIR FOR MOLD

Most of the time mold inspectors are called when a mold problem is suspected or water intrusion has occurred. Occasionally mold inspectors are called to simply check the air quality in a building in relation to mold. The inspector collects air samples and sends them to a laboratory for analysis. The process of testing air quality is called sampling. See *SAMPLING.*

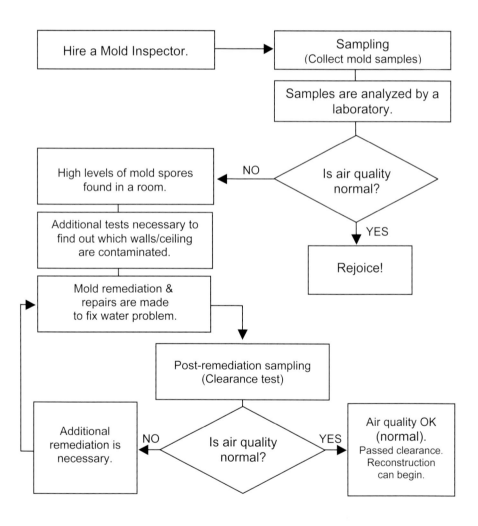

WHAT TYPE OF MOLD ASSESSMENT DO I NEED?

Many people are on a budget and are faced with the dilemma of choosing between what they need and what they can afford. If you are faced between the choice of a mold inspection and taking mold samples, always choose sampling. If you can afford doing both at the same time, do so. Below are some hypothetical situations. See *WHEN IS SAMPLING WARRANTED?* and *WHEN IS A MOLD INSPECTION WARRANTED?* The important question is "***What is your objective***?" If a renter needs to convince his landlord that there is visible mold in several rooms, a minimum of one surface sample will be sufficient to obtain a laboratory report to prove scientifically to the landlord that mold is present in the apartment.

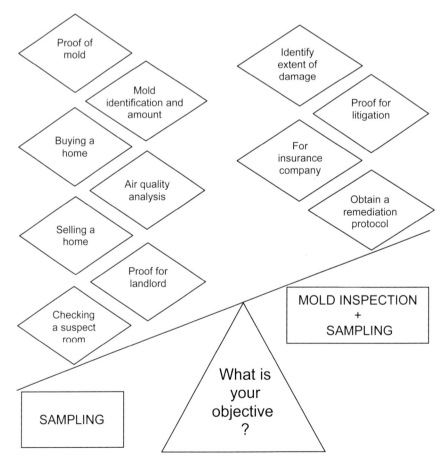

PROTECT YOURSELF

If water intrusion has taken place and building materials have not dried out within 48 hours, the chances of acquiring a mold problem increase exponentially. In the event of a natural disaster there are hordes cf unscrupulous wannabes who will descend into the area, set themselves up as "experts", and take advantage of unsuspecting homeowners. Be smart and heed the following 10 commandments:

10 COMMANDMENTS
TO AVOID GETTING RIPPED OFF

1. Thou shalt not hire fly-by-night companies.

2. Thou shalt not put your signature on a contract without reading the fine print.

3. Thou shalt not hire anyone without asking for references.

4. Thou shalt not skip the mold assessment phase.

5. Thou shalt not hire the same company to do both the finding (inspection) and the fixing (remediation).

6. Thou shalt not hire mold inspectors who analyze microbial samples in the back of their trucks.

7. Thou shalt not try to save a buck by hiring a handy man.

8. Thou shalt not skip the post-remediation sampling phase.

9. Thou shalt not trust miracle cures without scientific evidence that they have passed a clearance test.

10. Thou shalt not give full payment upfront to anyone.

REVELATIONS

The following tips are intended to help you become an informed consumer. Hopefully, they will help you avoid making the mistakes people we know have had to learn the hard way.

1. **THOU SHALT NOT HIRE FLY-BY-NIGHT COMPANIES.**
   - Make sure that the company you hire, whether it is a mold inspection or remediation company, is legitimate and intends to stay in business in your town. You need to see a physical local address on the contract, not just a P.O. Box. Disaster areas tend to draw self-made overnight experts who will be gone before you realize their incompetence.

   - The wannabe "Mold Inspector" might have just bought an air pump and learned how to operate the pump from the instruction booklet. He may not know any more than you do. If samples are not taken correctly the results will be worthless. If he does not know how to recognize red flags for mold, he will not be able to document the problem correctly and his report will not be accurate. He may not be able to find the source of moisture, which can sometimes be tricky even for the experienced mold inspector. Finally, he will not have a clue how to interpret the laboratory data, and will leave that important part up to you.

   - The wannabe "Mold Remediator" might leave you with a bigger mess than you had by turning a small problem into a major disaster. If he does not have the know-how to take necessary precautions and properly set up containment, he may contaminate your entire house. In many cases the work will have to be done all over again, and sometimes at more than twice the original cost.

2. **THOU SHALT NOT PUT YOUR SIGNATURE ON A CONTRACT WITHOUT READING THE FINE PRINT.**
    Read your mold remediation contract in its entirety. If you don't understand it or are unsure, don't sign it, have a lawyer look at it. A 15-minute lawyer consultation will save you money and prevent a lot of headaches. Be sure that the contract states the following conditions at a minimum:

    - SCOPE OF WORK REGARDING MOLD REMEDIATION: What the mold remediators are going to do, how they are going to do it, and what protocol they will follow.

    - RECONSTRUCTION (if applicable): What they are going to do, how they are going to do it, and the types of materials used. Some mold remediation companies do not do reconstruction.

    - SCHEDULE: A reasonable time schedule as to when they will start and when they anticipate the work to be completed.

    - POST-REMEDIATION SAMPLING (CLEARANCE TEST): Assurance that their work will pass air clearance (post-remediation sampling) performed by an independent inspection/testing company of your choice. In no instance should the remediators perform their own testing. There must be a contingency clause in the contract in the event the completed job does not pass air clearance the first time. In this case, it should state that they will remediate further at no additional cost until the work passes air clearance. It is also a good idea to spell out who will pay for the post-remediation sampling, and who will pay for the second post remediation sampling in the event the first test does not pass air clearance.

- HAULING AWAY TRASH: The hauling away of all contaminated materials from the premises.

- TOTAL COST OF REMEDIATION, AND TOTAL COST OF RECONSTRUCTION (If applicable). Full payment remitted after:
  - The work passes air clearance testing, and
  - The contaminated material is hauled away, and
  - Reconstruction is done to your satisfaction.

- CUSTOMARY CONTINGENCIES: Contingencies for breach of contract if any of the above conditions are not met.

3. THOU SHALT NOT HIRE ANYONE WITHOUT ASKING FOR REFERENCES.

- Legitimate companies will gladly give you references. The goal of a reputable company is to provide good service and make the customer happy so that, by word-of-mouth, their client base will grow and the company will be successful. Once you are given names and telephone numbers of former clients, call them and ask questions regarding the quality of workmanship, attitude of employees, and whether the time schedule estimate was reasonably accurate.

- Schedules are a very important aspect of remediation. Some companies cannot refuse business and they accept too many jobs. This means that they may have ten jobs going at once but they only have one crew. So you will see them one day and they'll be gone. On the 11th day you will see them again and you'll have to wait for your turn as they round robin the circuit of their customers.

- Ask what type of professional organizations they belong to, or whether they belong to the Better Business Bureau. If so, call the BBB and ask whether there have been any complaints about the company.

- Ask the mold inspection company for a sample copy of their reports. This will show you what to expect. The format will vary widely, from check-the-box to detailed narrative reports. Narrative reports are usually superior, but not always. This is where a professional is able to shine or to show his ignorance.

4. THOU SHALT NOT SKIP THE MOLD ASSESSMENT PHASE.

Unless the entire home is inundated with mold and the Inside needs to be gutted out, a mold inspector is necessary to assess the area in relation to mold, collect samples, and obtain a scientific report that will help qualify and quantify the problem. This is essential because without this phase mold remediators will design their own work.

5. THOU SHALT NOT HIRE THE SAME COMPANY TO DO THE FINDING AND THE FIXING.

A company offering to perform both services, mold detection/assessment and remediation, could lead to a conflict of interest. It is not appropriate for a mold remediation company to design their own work. Nor is it appropriate for them to do their own "clearance test", in effect checking their own work. You can be assured that they will "pass" their work with flying colors. You need to keep the processes separate. An independent mold inspector should perform the mold assessment, and a different company should perform the mold remediation. Neither of them should receive a "referral fee", percentage, or a monetary incentive of any kind from each other.

6. **THOU SHALT NOT HIRE MOLD INSPECTORS WHO ANALYZE THEIR OWN MICROBIAL SAMPLES IN THE BACK OF THEIR TRUCKS.**
    Don't laugh, some do.

7. **THOU SHALT NOT TRY TO SAVE A BUCK BY HIRING A HANDYMAN.**
    There are many instances when shopping for price makes sense. However, when dealing with mold you need quality work, not cheap work. In other words, you need a certified mold remediator.

    A mold remediator has been trained to mitigate a mold situation safely.

    ▪ Safely for himself and his crew by using protective clothing and masks.

    ▪ Safely for the occupants by setting up containment of the contaminated area. This is to prevent the billions of mold spores released by jostling contaminated drywall from spreading throughout the entire house.

    ▪ Safely by using the proper equipment. First, by creating a negative air pressure inside the containment area so that the air is sucked inward to prevent mold spores from leaving the area. Second, by having an air scrubber to remove particulates from the air as they work. Third, by changing the air scrubber's filters. Fourth, by sealing off the air scrubber openings (intake and outlet) when not in use.

- Safely so that the house does not get contaminated during the transportation of contaminated material to the outside for disposal. Everything removed from the area must be bagged in plastic prior to transporting it through the building.

8. **THOU SHALT NOT SKIP THE POST-REMEDIATION SAMPLING PHASE.**

This final stage should never be skipped no matter how "clean" the area may look. No-one can tell visually if the area has been successfully remediated; only laboratory results can be trusted to scientifically assess the air quality in relation to mold. If the results show an elevated amount of spores, the remediators must come back to further remediate the area.

9. **THOU SHALT NOT TRUST MIRACLE CURES WITHOUT SCIENTIFIC EVIDENCE THAT THEY PASSED A CLEARANCE TEST.**

Beware of new unproven methods and the latest miracle cure that promises to "kill" mold without traditional remediation. From sprays to fogging the entire house leaving a chemical film everywhere, these techniques are by and large unproven. And no, do not let them spray "herbicide" inside your house. Yes, we've heard that too.

10. **THOU SHALT NOT GIVE FULL PAYMENT UPFRONT TO ANYONE.**

Be smart, do not pay anyone in full upfront, and beware of scam artists. A reputable company will not ask you for full payment upfront. A deposit may be required at the start of work, but NEVER pay in full before the work is completed and has passed a final clearance, and you are fully satisfied

THE PROCESS OF MOLD REMOVAL

A 3-STEP PROCESS

1. THE MOLD ASSESSMENT PHASE

2. THE MOLD REMEDIATION PHASE

3. THE POST REMEDIATION PHASE

1. THE MOLD ASSESSMENT PHASE

The mold assessment phase comprises one or two services:

- Mold Inspection and/or

- Sampling

MOLD INSPECTION

WHAT IS A MOLD INSPECTION?

The objective of a mold inspection is to document in a report the condition of a home or building with regard to mold.

An inspection consists of a visual inspection of the surroundings with respect to "red flags", which can lead to mold growth. See the list *"WHAT ARE RED FLAGS"*. The inspector should test the walls for moisture using a moisture meter. An elevated reading is a red flag, because where there is moisture there is the possibility for mold growth. If a mold problem is suspected in a particular wall, a cavity sample should be taken, regardless of whether or not the wall tested dry. It is important to remember that If a leak happened at one time, such as during a hurricane, and the leak has since dried out, the moisture meter will read dry even though mold may be present on the inside of the wall.

A mold inspection can involve the entire building, inside and outside, only the inside, or a particular location within the building. For a single family home it may simply involve one room or floor. In the case of an apartment or condominium complex it may involve one or more apartments or condominiums, a single floor, or the entire building. It may involve checking the attic and/or basement. It is essential to find out what you are paying for, and the extent and limitations of the mold inspection.

If a mold problem is found, the inspector should also attempt to find and identify the source of water intrusion/leaks/moisture that led to the problem. If finding the moisture source is beyond his expertise then it may be necessary for him to recommend hiring a building scientist/engineer or other experts. Without identifying and repairing the moisture source before mold remediation takes place, mold will come back.

The mold inspector documents his findings in a Mold Inspection Report. If he finds red flags he should take photographs. Please note that not all mold inspectors provide photographs to their clients.

After completing his inspection, the inspector makes recommendations as to where samples ought to be taken. There should be a parallel between the list of red flags found and his sampling recommendations. In the event no red flags are found, an inspector should recommend a few samples to assess the indoor air quality in relation to mold. This is to make sure that an invisible mold problem is not missed, such as a problem inside the walls.

Depending on the findings of the inspection, he may recommend:
- Air samples to test the ambient air
- Wall or ceiling cavity sample(s) if walls or ceiling are suspected of having mold growth
- Surface samples to collect mold-like substances
- Carpet samples
- Bulk samples – a piece of drywall, carpet strip, etc.

A mold inspection report might be a check-the-box format or a narrative type format. A narrative report is preferable as they are usually more thorough and will normally follow the layout of the specific home or building. Regardless of the format, a narrative conclusion should be included, to recap the overall results, and a section for recommendations as well.

The Mold Inspection Report should state:
- Any red flags and their locations
- The relative humidity of the air
- The moisture content in walls if higher than normal
- Recommendation for mold samples
- The source of water intrusion/leaks/moisture if known
- General recommendations for mold remediation (not to be confused with a detailed mold remediation protocol)
- Suggestions if results were inconclusive

MOLD INSPECTION FLOWCHART

MLS= Mold-like substance

<u>Scenario</u>: Known leak. Visible MLS in some places + odor throughout the house.

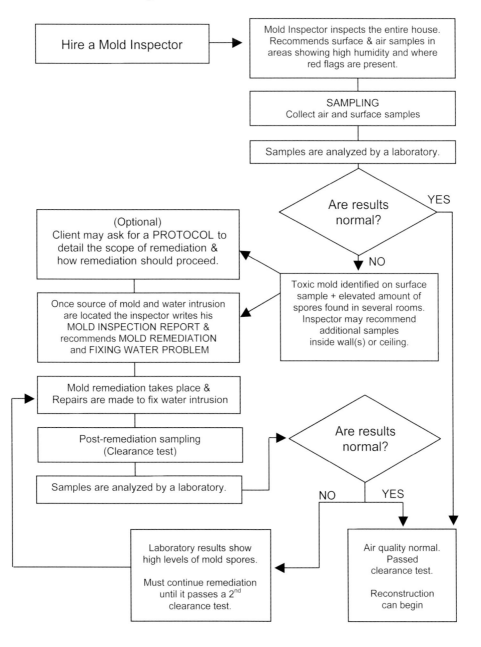

WHAT ARE RED FLAGS?

Red flags are conditions conducive to mold growth. MLS is a term used by mold inspectors to refer to a "mold-like substance" before it has been analyzed by a microbiologist.

INSIDE
- Musty smell
- Inside relative humidity above 60 percent
- Any visible water stains
- Any visible MLS (mold-like substance)
- Caulking missing around the bathtub
- Holes or grout missing in the shower
- Floor buckling or warping
- Leaky plumbing – under sink, toilet, washing machine, etc.
- Signs of corrosion, rusty nails
- Wood discoloration
- Paint discoloration, bubbling, or cracking
- MLS at air vents
- MLS in the air handler
- Two or more thermostats too close together
- A/C drain pan improperly tilted
- A/C drain line clogged
- Bathroom fans not working

OUTSIDE
- Blocked gutters
- Cracks in masonry, around windows
- Sprinkler heads too close to the house
- Sprinkler heads pointing towards the house
- Improper grading – slopes towards the house
- Built-in planter boxes
- Vines growing on the façade of the building
- Debris on the roof, leaves, vines, etc.
- Cracks in the chimney
- Any rotten wood on the outside of the structure
- Missing roof shingles
- Missing flashing
- EIFS – Can cause major problems if improperly installed.

PLEASE NOTE: The above list is not complete.

EXAMPLES OF RED FLAGS

Ferns growing
inside the
chimney
and
crack seen in the
chimney.

Vines and dead leaves

A leak   Sprinkler too close to the building

WHAT IS A MOLD REMEDIATION PROTOCOL?

A mold remediation protocol goes one step further than a Mold Assessment Report. It is a document that describes the scope of the mold remediation and is written following a mold inspection and after obtaining sampling results. It also addresses worker exposure to mold, worker protection requirements, work area cleaning, methods of removal, and waste disposal. Paramount in this protocol is the safety of the occupants and the workers, and preventing mold spores from leaving the contaminated area. Currently there are no Federal, State, or local regulations for evaluating the potential health effects of fungal contamination and remediation. However, reputable mold remediation companies should follow the guidelines for mold remediation as set forth by either the New York City Department of Health Guidelines for Mold Remediation, the Environmental Protection Agency Guidelines, or guidelines set forth by a professional organization to which they belong. See *THE MOLD REMEDIATION PHASE.*

A protocol is optional. Insurance companies often require a protocol from a mold inspector because it ensures that the mold remediators do not design their own work beyond what is required. Many mold remediation companies will not work without a protocol. It releases them from liability in the event that the scope of work is questioned down the road. Other insurance companies just need an inspection report and sampling results.

Building materials supporting fungal growth must be remediated as rapidly as possible in order to ensure a healthy environment. To prevent contamination from recurring, repair of the defects that led to water accumulation (or elevated humidity) should be performed prior to fungal remediation.

Extensive contamination should be remediated by personnel with training and experience in handling environmentally contaminated materials, particularly if heating/air conditioning (HVAC) systems, or large occupied spaces are involved. Effective communication with building occupants is an essential component of all remedial efforts. The use of respiratory protection, gloves, and eye protection is highly recommended.

All provisions stated in the protocol with regards to the health and safety of the workers, the public, and the protection of the environment are understood to be <u>minimum</u> <u>standards</u>. Furthermore, the full assessment of the damage must be left to the professional judgment and practicality of the remediation company once abatement provides full visibility of the mold-infested medium.

At the completion of remediation a mold inspector should be retained to conduct air sampling, also called "Post Remediation Sampling". It is always best to use the same inspection company and the same laboratory to obtain a direct comparison so keeping certain variables constant – the sample methodology during sample collection and laboratory analysis. Air clearance shall be achieved when indoor spore counts are less than the outside air, and both the indoor and outdoor samples consist primarily of the same types of mold.

WHEN IS A MOLD INSPECTION WARRANTED?

Performing a Mold Inspection is useful in the following situations:

- When the buyer or seller wants a visual inspection to document the condition of the property at the time of sale.

- When an insurance company, mortgage company, or attorney needs an inspection report to document the surroundings with respect to red flags.

- When a protocol is required. Prior to writing a protocol, the mold inspector needs to perform a mold inspection in order to determine the scope of work.

- When a homeowner or homebuyer wants to know if red flags are present that could lead to mold growth, so that corrective or proactive measures can be implemented. This may include replacing caulking around the bathtub, filling cracks in the stucco, moving sprinkler heads too close to the house, etc.

- When an elevated amount of spores is found in one room. The mold inspector can often perform an investigation of that room alone by testing the walls for moisture, looking for stains or color variations, looking at the baseboards, pulling the carpet to look at the carpet tack strips for rusty nails, and any other signs of water damage. Finding evidence of moisture will often lead the mold inspector to the mold colony.

WHAT TO LOOK FOR IN A MOLD INSPECTOR

A mold inspection should be conducted by someone who has received training by an accredited inspection school and who has passed a test that qualifies him as a mold inspector. Other experts in the field that may be consulted include an industrial hygienist, a forensic engineer, and an environmental specialist. As we mentioned earlier, a mold remediator should not perform mold inspections. Performing both functions, inspections and remediation, could create a conflict of interest - keep the finding (mold inspection) separate from the fixing (mold remediation). The inspector should also be licensed, insured, and bonded, and preferably have some experience.

A mold inspector should be able to perform the following services:

- **PERFORM A MOLD INSPECTION OR SURVEY**
  o Identify and document any red flags
  o Identify the source of water/moisture
  o Make recommendations with regards to mold remediation and repairs, if applicable
  o Write an understandable report

- **COLLECT MOLD SAMPLES, BOTH PRE AND POST REMEDIATION**
  o Collect mold samples using proper equipment
  o Apply the scientific method in his work
  o Follow sampling methodology
  o Keep an accurate sampling record
  o Provide a companion evaluation report with the lab report

- **WRITE A PROTOCOL BASED UPON:**
  o A visual mold inspection
  o The laboratory sample analysis
  o Recognized professional standards for mold remediation.

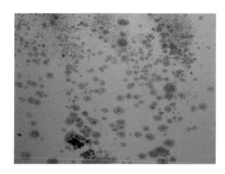

In homes with serious mold problems the mold inspector suits up to protect his health.

A mold inspector should take the time to listen to you. In so doing he will gather a wealth of information. You are in the perfect position to recount the sequence of events, what you have seen, or what you suspect. The information gathered will greatly help the mold inspector to develop a proper strategy in his search for mold and the source of water intrusion. Depending on the situation and the final objective he may recommend any of the following:

- Collect mold samples in one or more rooms
- Collect air sample(s) in one or more wall/ceiling cavities
- Collect samples throughout the entire building
- Inspect one area only – room, attic, basement
- Inspect the entire building inside
- Inspect the entire building inside and outside
- Inspect, collect samples, and write a protocol

He should take the time to write a comprehensive report in layman's terms to explain the results and to answer any further questions you might have.

As part of his mold inspection, the inspector should try to locate the source of water or moisture that caused mold to grow. Unless the water/moisture problem is found and fixed before mold remediation, mold will recur. Building construction is very complex, and as such the task of the mold inspector can be difficult. The building design, the choice of material and human errors, before and during construction, are just a few of the problems that he might encounter. There may be an occasion when finding the source of moisture will be beyond his expertise. He should then recommend to his client to hire a forensic or structural engineer.

Finally, a good mold inspector enjoys and takes pride in his trade and continues to learn about mold by taking various classes to improve his techniques and expertise.

WHAT TO EXPECT IN A
MOLD INSPECTION REPORT

A mold inspection report details the visual inspection just performed. Besides describing red flags and/or locations of MLS, a narrative conclusion should be written to recapitulate the overall results, as well as to include recommendations, if applicable.

The report should state:

- **RED FLAGS**
  - Their locations
  - Repair recommendations

- **MOLD-LIKE SUBSTANCES (MLS)**
  - Their locations
  - Amounts visible (subjective) – low, moderate, high

- **MOISTURE CONTENT**
  - Assess moisture content in walls
  - Evaluate the relative humidity of the air

- **SOURCE OF WATER/MOISTURE/LEAKS**
  - Identify the location
  - Recommendations for repairs

- **SAMPLING**
  - Sampling was conducted prior to the inspection and is deemed sufficient per results of the inspection - no further sampling is needed.
  - Sampling was conducted prior to the inspection and other suspect areas call for additional sampling.
  - No prior samples taken – inspector recommends sampling to assess the indoor air quality in relation to mold, regardless of the overall visual conditions of the home or building.

- **CONCLUSION**
  - o Recap the visual mold inspection

- **RECOMMENDATIONS**
  - o If applicable, general recommendations for mold remediation if prior sampling revealed or confirmed a mold problem. General recommendations for mold remediation and repairs are not to be confused with a detailed mold remediation protocol. Other recommendations may include hiring the expertise of a roofing company or of an HVAC expert.

WHAT EQUIPMENT
IS USED IN MOLD / WATER DETECTION?

Certain instruments are excellent in mold or moisture detection. However it is important to understand their benefits and limitations.

MOISTURE METER

Photo courtesy of GE Sensing, Inc.

A moisture meter is the most common instrument used by both home and mold inspectors. This tool is useful to detect moisture in different mediums. Usually, a moisture meter is designed to read moisture content in wood and the wood moisture equivalent (WME) in other building materials – the WME system of measurement means that other materials, such as drywall, plaster, flooring, and concrete can be measured using a moisture meter calibrated for wood.

There are two types of moisture meters, those that use non-invasive RF technology and those that use pin type resistance measurement. Non-invasive meters are extremely valuable in scanning drywall, floors, and wall coverings such as tile and vinyl. Once an elevated reading is detected, the pins can be used to obtain more accurate readings. For this reason, a meter that uses both measurement principles is favored.

The moisture meter is an excellent instrument to test moisture when an active leak is present, or the leak is fairly recent. When one or more walls show high moisture readings, a mold sample should be taken in those walls to determine if mold is present.

The novice mold inspector will swear by his moisture meter, while the experienced mold inspector knows its benefits but also its limitations. If the walls test "dry" it does not mean there is no mold inside the wall cavity. Consider the following. If water intrusion occurs and mold begins to grow on the inside of the wall cavity, it will remain there whether or not the source of moisture is fixed. Over time the wall will dry out, but sometimes you will get lucky; mold may not have grown at all. It all depends on the amount of water that came in and the time it took to dry out. The longer it took to dry, the higher the probability that mold has settled in and grown. This means that if there is a known leak, old or new, samples should be taken to assess the wall cavity.

DIGITAL CAMERA

Photo courtesy of
Olympus America, Inc.

A mold inspector should take digital pictures of red flags and put them on CD as part of his mold inspection. Not all inspectors provide this service. In addition, a mold inspector should take pictures for quality control and have a record in the event his sampling methodology is ever questioned. In the picture he should have a placard showing the Client ID, date, sample ID, and sample location. His pictures should show the air pump in relation to the surroundings, the actual canister showing the sample ID number, and the flow rate of the pump. Pictures are worth a thousand words, and millions of dollars to lawyers. Examples of quality control procedures and pictures are seen in the following pages.

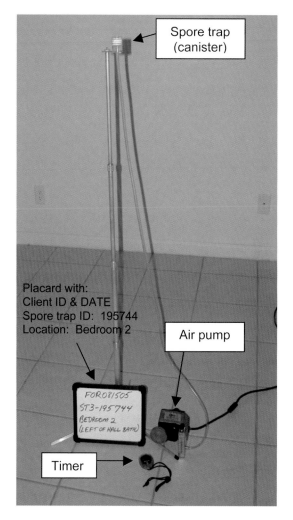

Spore trap (canister)

Placard with:
Client ID & DATE
Spore trap ID: 195744
Location: Bedroom 2

Air pump

FOR081505
ST3-195744
BEDROOM 2
(LEFT OF HALL BATH)

Timer

AIR PUMP

An air pump is used to take air samples. A spore trap (canister) is inserted at the end of a tube, which in turn is attached to the pump. Inside the canister is a slide with a sticky substance. A precise volume of air is pumped through the spore trap. The volume of air sampled is based upon airflow rate and the length of time pumped. Once the pump is turned on the mold spores floating in the air are captured on the slide.

When the sample arrives at the laboratory, the microbiologist breaks the seal, retrieves the slide, and puts it under the microscope for analysis. His analysis consists of identifying the types (genera) of mold spores and counting the number of spores found for each type of mold. With the known volume of air, he can then convert the actual number of spores on the slide to a standard unit of measurement – spores per cubic meter.

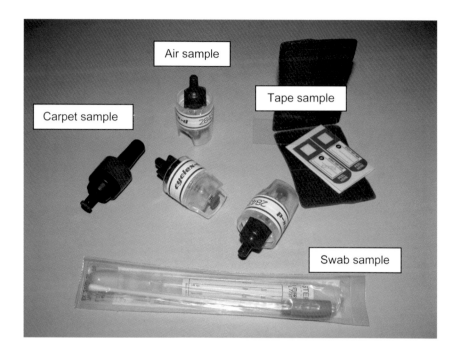

SAMPLES

Various types of samples are used to test the environment for mold.

- Surface samples, such as a tape or swab, are used to test visible mold-like substances.

- Air samples are taken using spore traps (canisters) to collect mold spores floating in the air. A baseline sample is collected outside as a control.

- Carpet samples are used to collect mold spores found in carpets.

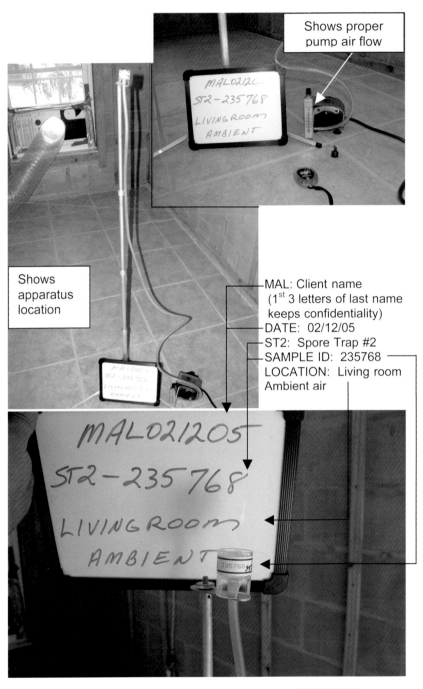

Shows proper
pump air flow

Shows
apparatus
location

MAL: Client name
(1st 3 letters of last name
keeps confidentiality)
DATE: 02/12/05
ST2: Spore Trap #2
SAMPLE ID: 235768
LOCATION: Living room
Ambient air

POST-REMEDIATION SAMPLING PICTURES

Quality Control At The Job Site

MICROSCOPE

Specialized microscope used by microbiologists in a laboratory setting to analyze mold samples. The microbiologists are able to qualify and quantify the samples by identifying the genera and species of mold and their respective amounts.

Photo courtesy of
Olympus America, Inc.

MOLD DOGS

Some mold inspectors use dogs that have been properly trained to detect mold. Their excellent sense of smell helps them pinpoint a mold-contaminated area. Once a dog has discovered a mold source, the mold inspector will take air samples to qualify and quantify the contamination. Mold dogs are said to be tools. However, unlike other types of equipment there are controversial ethical issues in using man's best friend as a tool knowing that animals, like humans, can get sick from mold. Some dogs develop *nasal Aspergillosis*, or cancer. A mold inspector can protect himself with an appropriate mask and gear, but the dog lacks total protection since his sense of smell is used as a tool.

FIBER OPTIC INSTRUMENTS

Specialized instruments allow the inspector to see inside walls with little difficulty. Slim flexible cameras and fiber optics fit into small spaces and through small holes to view the inner wall cavities. With the help of a fiber optic instrument the mold inspector can view hard to reach areas to help locate mold growth and obtain an idea of the amount mold contamination. However, they have limitations. They allow the inspector to see only a few feet all around; this means that several holes may need to be drilled into the wall if nothing was found on the first try. The holes range from pencil sized to 1.5 inches in diameter depending on the brand of instrument. Stud partitions and Insulation can limit the use of fiber optic instruments.

PARTICLE COUNTERS

Particle counters give a count of particulates present in the air. They have limitations in mold detection applications. Each counter gives a count according to the size of particulates. Particulates are everything floating in the air, including skin cells. Different counters have different ranges. In case of mold spores, the size varies widely, so having a count of particulates, let's say between 1 and 3 microns, would not tell us much, because there is no way of knowing what they are. Some mold inspectors use a particle counter for post-remediation testing. If the count is relatively low, they pass the remediation. This is inappropriate. Air samples should be taken to obtain a qualifying and quantifying test - learn what genus of mold spores are present and their respective amounts. Particle counters can be useful to help locate a room with higher concentration of particulates than another. In this instance they help narrow the search when a mold contamination is suspected and no red flags are present.

INFRARED CAMERA

An infrared survey can be performed to determine the extent of water damage. Thermal infrared radiation (IR) is emitted by all objects whose temperatures are above Absolute Zero. Infrared is simply light (electromagnetic) waves that are invisible to the eye, but that can be felt as heat. Infrared Thermography is a technology that has produced devices that allow these "invisible" heat waves to be seen and captured on video, film, etc. Thermal imaging equipment shows objects having colors that correspond to their temperatures. These images allow any unusual heating or cooling (hence, problems that may be starting) to be seen, even if it is not yet visible to the naked eye. Similarly when an object is heated to a high enough temperature, it will start to glow in the visible portion of the electromagnetic spectrum and will be seen as red or white-hot.

An infrared camera helps detect moisture in building materials when an <u>active</u> or <u>recent leak</u> is present. Over time, the moisture in building materials dry out and one can no longer detect any anomalies, even though sample results may show high levels of mold spores behind walls and ceilings. Infrared inspection does not directly detect the presence of mold; rather, it is used to find moisture where mold may develop.

Thermal Imaging is the avant-garde tool of predictive maintenance that helps avoid costly repairs, temporary breakdown, production interruption, loss of time, sickness or death, and/or potential lawsuits. By doing periodic audits of structures and systems, potential problems can be averted.

Large companies are becoming proactive in taking preventive measures through periodic thermal imaging of their facilities or plants. Private individuals are also taking advantage of this unique predictive safety maintenance tool to protect their own personal real estate.

If a mold inspector claims that he can perform an infrared survey of your home or building and tell you whether there is mold growing anywhere, then don't hire him. He knows nothing about mold. See paper at the end of this book, titled *"PLAYING HIDE AND SEEK WITH MOLD"*.

SAMPLING

WHAT IS SAMPLING?

If you ever thought about purchasing a do-it-yourself mold test kit, don't. A mold kit is basically a Petri dish. "Petri" comes from a German bacteriologist, Julius Petri, who came up with the technique of using little dishes with a perfect environment for growing bacterial strains in his laboratory. You learned earlier that mold spores are everywhere and when they find a good environment they settle and grow. By setting a Petri dish on a table you are betting that mold spores may or may not land on the dish. Although interesting, it will give you no useful information. Sampling is a scientific procedure; testing air quality cannot be done with a do-it-yourself kit.

There are two main types of sampling methods: Culturable and non-culturable samples. The culturable method allows the microbiologist to differentiate between species whose spores are visually similar, but it takes longer to obtain the results and it is also more costly. In addition, there are certain types of mold that do not grow well in a laboratory setting. Mold spores collected in a residence are grown (cultured) in a laboratory and analyzed after more extensive growth has occurred. Thus, it is believed that the non-culturable method provides a more accurate "snapshot" of the air besides being cheaper and quicker. The method outlined in this book refers to the spore trap (non-culturable) method.[18]

Sampling is the scientific approach to mold assessment, the purpose of which is to qualify and quantify the environment in relation to mold. According to the American Industrial Hygiene Association (AIHA) about 50% of mold problems are not visible. Thus, sampling is a vital tool to help assess air quality even when mold is not visible.[19]

Certain conditions must be met before sampling takes place. For 48 hours prior to sampling any stand-alone air cleaning devices must be turned off and all windows and doors must remain closed. Normal traffic in and out of the house or building is ok. The reason for this is to obtain a true reading on indoor air quality without any filtering devices or mixing indoor/outdoor air. In addition, the inspector must wait at least two hours after the last rain to collect an outdoor control air sample. This is because during rain, the spores floating in the air fall to the ground. If the above conditions are not met, the sampling results would be inaccurate.

The mold inspector should follow a rigid sampling protocol. In order to obtain accurate results, samples must be collected in a precise and controlled manner by a mold inspector in the field and then analyzed with a high degree of precision by a degreed microbiologist in a laboratory.

The mold inspector should keep both a written and visual record of the sampling as it is performed. He should also provide you with a copy of the sampling records at the time of sampling, and reference them again in his report. Should anyone question the results, the mold inspector ought to be able to provide proof of his sampling methodology. Likewise, the laboratory should follow a precise methodology and keep impeccable records.

You should receive two reports; a laboratory report and a companion report by the mold inspector. We call our companion report a *Mold Assessment Report*. The latter should explain the laboratory results in more detail and provide recommendations, if necessary.

Sample results provide a snapshot of the environment at the time of sampling at a particular location, as conditions can change over time. Mold can start growing within 24 to 48 hours when food and a moisture source are present. The results do not guarantee that conditions will not change or that mold may/may not grow on the premises in the future.

In general, a mold problem exists when the amount of mold spores found inside a home/building is greater than or different from the types and levels found outside. For more details, see *HOW TO UNDERSTAND YOUR LABORATORY REPORT*.

SAMPLING FLOWCHART

MLS= Mold-like substance

<u>Scenario</u>: A <u>home</u> inspector noted in his report that he saw something that looked like mold in the air handler. Sellers disclosed having had mold on the north wall of the dining room, but stated that it was remediated and the leak was repaired. Homebuyer wants to have the home checked for mold by sampling.

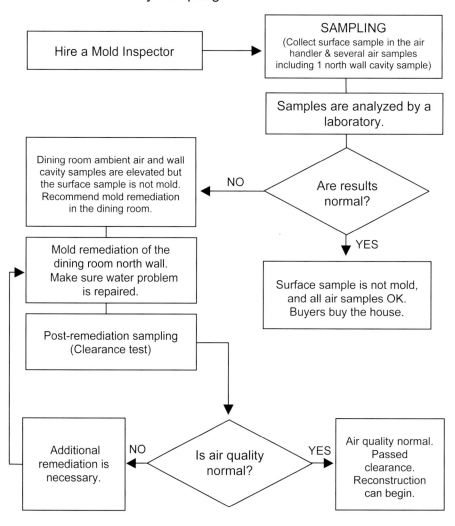

WHEN IS SAMPLING WARRANTED?

"Hello, I need to make an appointment for a Mold Inspection".

This is the first thing we hear from callers. Once we start asking questions, 95% of the time we will recommend sampling first. Once sampling is performed, we will end up investigating only a handful of cases because, very often, a mold inspection is not needed.

We recognize that every situation is different and there are times when we will recommend an inspection first, but most of the time sampling should be done first.

The following are examples of situations where sampling is warranted.

PROBLEM KNOWN OR SUSPECTED

- When a <u>mold</u> inspector recommends a certain number of samples in specific places based upon red flags found during a mold inspection.

- When a <u>home/building</u> inspector has found an elevated level of moisture in one wall and he recommends having the wall or room checked by a mold inspector.

- When a suspect area has already been identified by the owner, sampling will determine whether a mold problem exists.

- When an initial sampling showed an elevated amount of mold spores in one room and additional samples are needed to help find the culprit wall or ceiling.

PROBLEM UNKNOWN

- When a prospective buyer wants to know if there is mold growing in the building, and the home inspector did not find anything unusual, we recommend sampling instead of a formal mold inspection. Find out the reasons in the section *REAL ESTATE TRANSACTION TIPS*.

- When a family member is suddenly developing allergies and you want to determine whether it could be caused by mold growing somewhere in the house, sampling will provide you with a scientific mold assessment regarding the air quality.

- When several rooms have had water intrusion and no visible mold is present, sampling will qualify and quantify the air quality in relation to mold and identify the rooms with elevated amounts of mold spores while exonerating other rooms.

- When the owner of a company wants his building checked for mold because of complaints by his employees.

- When the owner wants testing performed for prevention and peace of mind.

WHAT TO EXPECT IN
A LABORATORY REPORT

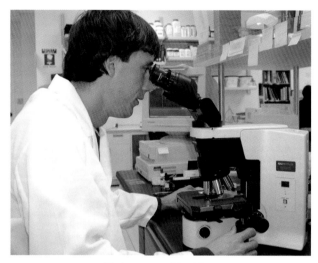

Picture provided by:
Environmental Microbiology Laboratory, Inc.

Sample analysis should be performed by an accredited laboratory that has also received AIHA EMLAP accreditation. Participation in the Environmental Microbiology Laboratory Accreditation Program is voluntary and a mark of distinction. It insures quality analyses.

The laboratory report should be clear and understandable and should provide you with an interpretation of the results based upon:

- Qualitative analysis – genus and species identification
- Quantitative analysis – quantity of spores compared to the outside control
- Commonality or rarity of the mold species
- Known or suspected toxicity

WHAT TO EXPECT IN A SAMPLING REPORT

In his report, the mold inspector may refer to mold-like substance (MLS) to describe what appears to him to be visible mold. This appellation is necessary because mold is a microscopic organism and, technically, no one can call it mold until it has been sampled and identified by a laboratory. The Mold Assessment Report is a companion to the Laboratory Report and is based on the data obtained from the laboratory. It is written by the mold inspector in layman's terms, and should include the following:

- **SAMPLING RECORD**
  - Sample ID
  - Identification of the sampling location
  - Total volume of sample (in Liters)

- **COMMENTARY ON THE LABORATORY RESULTS**
  - The amount of spores found at each location
  - The types of mold found
  - Explanation of the types of mold found based upon the commonality or rarity of the species

- **CONCLUSION**
  This should clearly state whether the findings were within the normal range. If not, it should state the degree and location of any mold contamination.

- **RECOMMENDATIONS**
  - General recommendation for mold remediation (not to be confused with a detailed mold remediation protocol).
  - Recommendation for additional samples, should the results be inconclusive.
  - Recommendation for additional samples to help locate the source of mold (if necessary).

WHAT KIND OF SAMPLES DO I NEED?

There are different kinds of samples for different situations. You must rely on your mold inspector to advise you, and to answer any questions you might have.

SURFACE SAMPLE

- If mold-like substance is visible, the inspector should collect a surface sample in addition to an air sample. A tape-lift or swab is used to collect the sample. Occasionally no mold growth is found on the sample - this means that the MLS was actually soot, or dirt, or another substance. If mold is present the laboratory will identify the species of mold. However, it will not show the quantity. The amount found is related to the amount on the swab or tape-lift, which does not give the extent of mold contamination on the wall. Essentially, a surface sample has no control or baseline to which it can be compared.

  A surface sample is qualifying – it helps identify precisely the <u>species of mold</u> found on the sample.

- As a rule, surface samples complement the results of air samples. Certain types of mold, such as *Stachybotrys*, do not spore easily. For this reason, it is possible that an ambient air sample may not pick up these mold spores, even though a mold colony may be present.

AIR SAMPLE

- Air samples are the most common specimen collected. To test the air quality of a home or building, ambient air samples and a corresponding outside control are collected using an air pump. Inside each spore trap (canister) is a slide with a sticky substance. As air is pumped through the canister, the mold spores stick to the slide. The mold inspector is actually collecting the mold spores that are floating in the air. A microbiologist in the laboratory will identify and count the different mold spores collected on the slide and compare the results to the genera and amounts found in the outside control.

An air sample is both qualifying and quantifying – it helps identify the types of mold spores and tells us how many mold spores of each genus is present compared to the outside control.

▪ If mold is growing behind drywall, the spores may or may not have yet percolated through the wall. For reasons not yet understood the proportional amount of spores coming through the ambient air of the room is not consistent from building to building. It is not unusual for an ambient air sample of a room to be within the normal range while a high level of mold spores is found inside the wall cavity. We suspect the house pressure (negative or positive) to be responsible for this phenomenon. Other possible explanations could be the type of sheetrock, or the choice of sealer and paint.

▪ DIFFERENCES BETWEEN AN AMBIENT AIR SAMPLE AND A WALL/CEILING CAVITY SAMPLE
   o An ambient air sample helps determine the air quality inside the room compared to one outside control in terms of genera and the mold spore levels (amounts). It tells what the occupants are breathing in terms of their exposure to mold.

   o A wall cavity or ceiling sample is used when mold is suspected behind the sheetrock. The inspector can insert a tube attached to the air pump through an electrical outlet or, with the permission of the owner, drill a pencil-size hole to insert the tube through the sheetrock.

- **CARPET SAMPLE**
    - A carpet sample is collected to test for mold when water is known to have soaked the carpet. An area of approximately 100 cm^2 is vacuumed using an air pump. The canister used appears similar to an air sample spore trap; however, a paper filter traps particles vacuumed by the pump.

- **BULK SAMPLE**
    - A piece of organic material showing evidence of a mold-like substance is cut out, double bagged in plastic, and sent to the laboratory for analysis.

HOW MANY SAMPLES DO I NEED?

The number of samples depends on the needs and objectives of the client (owner, buyer, mortgage company, or lawyer), and whether the problem is known or unknown.

Every mold contamination problem is different. Once on site the inspector will be able to advise you as to the number and types of samples that are needed for your particular situation. It is important to know that the accuracy of sampling increases with the number of samples collected. If mold is suspected inside the wall/ceiling cavity, we recommend that additional samples be taken in that wall cavity. Ultimately, you have to give your authorization to the mold inspector on both the manner of sampling and the total number of samples to be taken.

It is always easier to prove a positive – the presence of a mold problem, than to prove a negative – no mold problem exists. There are many factors that influence the accuracy of a sample and that can produce a false negative. In other words the sample results could show that the levels are within the normal range when in fact there is a mold problem somewhere. Some factors are totally beyond the control of the mold inspector. See *WHAT ARE THE FACTORS THAT PRODUCE FALSE NEGATIVES.*

Over the years we have developed our own sampling guidelines to determine the number of samples for a home/building, depending on whether the problems are known or unknown.

OUR SAMPLING GUIDELINES

The sampling guidelines outlined in this section are understood to be a <u>minimum</u>. The number of samples we recommend depends upon:
1. The size and other characteristics of the building i.e., one sample per 300 sq. ft., the floor plan, individual rooms, and number of stories; and,
2. Whether there are known problem areas or unknown problems.

The cost of sampling enters in the decision process because the more samples, the higher the cost. The goal is to collect an adequate number of samples to achieve high accuracy, while still keeping it affordable to the client.

The number of samples depends on whether the mold inspector is testing a known problem area as evidenced by red flags, or whether he is testing an entire home or building with no suspected problems.

Since every building layout is different the recommended number of samples and their types ultimately must be left to the professional judgment of the mold inspector.

SAMPLING A KNOWN PROBLEM AREA

Observation: Any red flags; see *WHAT ARE RED FLAGS?*
Objective: Assess the air quality of the area(s) of concern to qualify and quantity the mold problem.

Taking ambient air samples will show what the occupants are breathing. It will help qualify and quantify the problem, and it may help discover other unseen mold problems.

The sampling location should include the problem area(s), nearby non-problem area(s), and one outside sample to be used as a control for data interpretation. The laboratory results obtained from the non-problem area can help locate extended contamination from the first area or help discover a possible secondary problem.

If mold is suspected behind a wall, an air sample should also be collected within the wall cavity. A tube can be inserted through an electrical outlet box or by drilling a pencil-size hole into the drywall. To provide mechanical agitation the inspector should slap the wall and wait a few minutes before collecting the sample. Should a mold colony exist inside the wall the agitation will help spores become airborne, especially in the case of molds, such as *Stachybotrys*, which does not aerosolize easily. If a hole is drilled in the wall, the inspector should seal it immediately with tape after collecting the sample.

If mold-like substance is visible, the inspector should also collect a surface, carpet, or bulk sample(s) in addition to air samples.

SAMPLING FOR UNKNOWN PROBLEMS

Observation: Nothing visible
Objective: Assess the air quality of a home or building

Air samples are collected. The goal is to collect enough samples at strategic locations to provide adequate coverage plus one outside control. Should a mold problem exist in the building the results of the samples will flag the general area of the mold contamination. Once a problem is found a mold investigation can be performed and/or additional samples collected to locate the mold colony.

- **SINGLE STORY DWELLING**
  Home, Apartment, Condo, Town home
  For a 1,200 sq. ft. home imagine a square. One sample is collected at each corner. The number of samples is modified according to the floor plan, i.e. rectangular, L-

shaped, or U-shaped so that proper coverage is achieved.

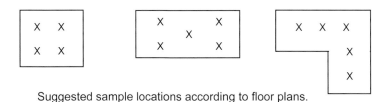

Suggested sample locations according to floor plans.

DWELLING WITH TWO OR MORE STORIES
Same as above multiplied by the number of floors.

- ### APARTMENT COMPLEX
Same as single story dwelling multiplied by the number of units.

- ### COMMERCIAL OR INDUSTRIAL
One (1) air sample per office, or one (1) air sample per 300 sq. ft. For a large open floor plan with cubicle type offices, the square footage is used to determine the number of samples.

For purchase of a large complex with several buildings, we recommend collecting samples in randomly chosen units in each building. Generally sampling 20% of the units in each building will provide a good idea of the overall condition of the complex.

OUTSIDE SAMPLE
As a rule one (1) outside air sample is sufficient to be compared with the results obtained with inside samples. For multiple stories, one (1) outside sample is collected at floor level on a balcony. However, when sampling is performed for litigation purposes, it is recommended that several outside air samples be collected and the results averaged to obtain a more representative sample of the outdoor environment.

WHAT ARE FACTORS THAT PRODUCE FALSE NEGATIVES?

There are various factors that produce false negatives. This means that sample results indicate that the mold spore levels are within the normal range compared to an outside control, when in fact there is a mold problem somewhere. Again, we want to stress that mold detection is not an exact science. This is when expertise may make a difference in avoiding false negatives. The mold inspector must keep in mind all the following factors when recommending the number of samples and their locations:

- **BAD TIMING**
  For some reason, the mold is not sporing at the time of sampling. The mold is too recent - mold spores may not have had a chance to percolate through the wall and permeate the ambient air inside the room.

- **NO AIR MOVEMENT**
  Wall cavities and the space between two floors provide little to no air movement. Some molds require an active disturbance in order to spore. If mold is present a mechanical agitation, such as a tapping on the wall prior to sampling, may help spores go airborne.

- **SOME MOLDS DO NOT SPORE EASILY**
  Some molds, such as *Stachybotrys*, do not spore easily. It is possible to find a visible mold colony of *Stachybotrys* on the wall with no mold spores of *Stachybotrys* in an air sample taken in the same room.

- **WRONG PLACE**
  The wall cavity sample is collected in an isolated partition between two studs with heavy insulation but a mold problem exists in another partition.

- **DILUTED AIR**
  Large rooms, such as a large office space can dilute the levels of mold spores because of the large volume of air involved.

- **POSITIVE PRESSURE**
  If the house is under negative pressure, mold spores found inside wall cavities will be pushed inside the rooms. Ideally, the house should be in a slight positive pressure. If a house is under strong positive pressure, that is, the air is being pushed towards the outside wall, any mold spores in the outside wall cavities will be far less likely to percolate towards the inside of the rooms.

- **HOT AIR RISES**
  Mold spores between two floors will tend to stay in the cavity because hot air rises and the cavity is usually warmer than the room below. We have observed a high level of mold inside a ceiling with a large hole while an air sample taken in the room below showed normal levels of mold. The air did not mix.

- **BUDGET CONSTRAINTS**
  Budget constraints can sometimes jeopardize mold sampling accuracy. The mold inspector recommends a certain number of samples based upon a combination of the number of rooms, the square footage, the floor plan, and his visual observation. As a rule the mold inspector does not recommend wall cavity samples unless he suspects a mold problem within the wall. The more samples are taken, the greater the accuracy, but the greater the cost.

  The budget of the client will ultimately dictates how many samples will be taken regardless of the inspector's advice. The competent inspector has to suggest a minimum and adequate number of samples without jeopardizing accuracy. His expertise and honesty can sometimes result in losing a project in the bidding stage to a novice or unscrupulous competitor who may recommend an insufficient number of samples.

Not all inspections and sampling are equal in their scope. When comparing several bids always look at the type of service you are getting and the number of samples recommended, not just the total price. Remember, a low bid often means low quality.

- **AIR CLEANERS ON, WINDOWS OPEN**
  We warn our clients that for 48 hours prior to sampling, the windows must stay shut and any air cleaning devises must be turned off. We have encountered several situations where wannabe mold inspectors took samples in homes (with mold problems) while air cleaners were on or the windows were open. It was not surprising that the results were normal!

HOW TO UNDERSTAND
YOUR LABORATORY REPORT

Laboratories convert the mold spore count found in the sample into mold spores per cubic meter. This type of measurement helps standardize the quantity for comparison purposes.

The types of spores and their respective amounts found outdoors on a given day vary according to geographical location, the vegetation (plants and trees) found nearby, and the ever changing daily weather. If two sets of outdoor samples have been collected on two different days, it is likely that the outdoor spore count will be different. The amount of spores found inside does not fluctuate as rapidly as the outside, as the levels are a function of traffic in and out of the residence or building and the amount of mold growing inside.

There are different brands of spore traps. Brand A may collect more spores than brand B. This does not mean that brand A is better than B because, in air samples, all spore amounts are relative to the inside versus outside spore count. What is important is the ratio between the number of spores found inside versus the number of spores found outside.

It is important that the microbiologist is able to "read" the slide without having too much background debris on the slide, which could hide the spores and affect the results. Therefore the laboratory will advise the mold inspector on the volume of air that should be collected for a particular spore trap. That volume of air will then be taken into account when converting to the amount of spores per cubic meter.

The interpretation of the results should preferably be done by two people: the microbiologist and the mold inspector. Some laboratories provide sample results (spore counts), without any interpretation. It is always preferable to have a laboratory provide an interpretation of the data, not just numbers. Ultimately it is the responsibility of the mold inspector to provide the client with a report that explains the data in detail with recommendations whenever applicable.

At the moment there are no federal and few state guidelines regarding appropriate levels for assessing mold contamination, or to determine whether certain levels are acceptable. Because people's sensitivity to mold differs widely and because certain segments of the population are more affected than others, it would be extremely difficult to come up with a specific numerical value. Geographical location and daily weather affect outside mold spore types and amounts. These factors impede arriving at a constant baseline to which inside samples can otherwise be compared.

The laboratory results are interpreted based on the scientific data obtained from inside/outside samples taken at <u>a particular location</u> on a <u>specific day</u>. The interpretation is based on the types of fungi present, the spore counts, ratios, the commonality and rarity of spores, and the degree of known or suspected toxicity.

The outside sample provides a baseline to which the indoor samples can be compared <u>for each individual type of mold</u>. The amount of spores found inside a home/building should be lower than the amount found outside, and the genera and their ratios be comparable.

There are four types of mold spores that are commonly found outdoors. It is natural that these common molds will also be found inside a home/building in larger amounts than rare molds. Molds found in abundance in nature are:
- *Penicillium*
- *Aspergillus*
- *Cladosporium*
- *Basidiospores*

If higher amounts of mold spores from these species are found inside it indicates that a mold source is likely present indoors. Even relatively benign molds found in large quantities may cause health problems.

Basidiospores are related to mushrooms and fluctuate and spike wildly after a rainy day. This is a normal occurrence. However, when found inside in large amounts it can be an indication of wood rot.

Stachybotrys and *Chaetomium* spores are generally found in extremely small amounts outdoors. Consequently their presence inside, even in relatively low numbers, is often an indication that they originated from mold growing indoors. *Stachybotrys* mold spores are not easily airborne. When a low amount of *Stachybotrys* spores is found, the severity of contamination cannot be assessed by spore counts alone. In this case further investigation will be warranted to locate the mold colony.

Stachybotrys is usually found in an environment that has been wet for a long period of time. *Chaetomium* is found on a variety of substrates containing cellulose including paper and plant compost. It has been found on paper in sheetrock that is, or has been, wet. Both have the potential of producing toxins. While interpreting the results, the amounts of these rare spores naturally carries more weight than spores that are found in abundance in nature.

Some mold inspectors will only compare the total amount of spores found inside to the total amount found outside. This is a grave mistake. Each type of mold should be evaluated and compared separately. It is possible to find the total amount of spores inside to be lower than the outside while a serious mold problem is present. See *AIR SAMPLE RESULT INTERPRETATION*, example #2, which shows a serious mold problem with *Stachybotrys* mold spores even though the total number of mold spores found inside is lower than those found outside.

If the laboratory results show a normal level for all species of mold found in that sample, one can say that the air quality is within the normal range compared to the outside air. In the event that the sample results taken in the ambient air of a room are normal while a sample taken inside the wall cavity shows a high level of mold, we can say that a mold problem exists inside the wall cavity and the wall needs to be remediated.

Wall cavity samples have to be carefully interpreted because there are many factors that can influence the results. The volume of air inside a wall/ceiling cavity is small compared to the ambient air of a room. There is also a total lack of visibility as to the distance between a possible mold source and the sample tube, i.e. a mold colony could be 10 feet away or be 10 inches away. Due to these facts, the results should be carefully evaluated to determine the severity of contamination inside a wall or ceiling cavity. Nevertheless, taking a cavity sample is still the best way to determine if a mold problem exists inside a wall or ceiling. Your mold inspector should interpret the laboratory results of wall cavities for you.

As explained earlier it is best to choose a laboratory that interprets their results. However, the microbiologist in the laboratory only sees what is in the samples, while the inspector sees the overall conditions in the building and knows additional facts of the situation. Once the laboratory results are known to the mold inspector, he may have to make a judgment call to investigate further and/or to take additional samples if the results were inconclusive. Ultimately it is your mold inspector who should convey and explain the laboratory findings in a report and make his recommendations.

AIR SAMPLE RESULT INTERPRETATION EXAMPLES

EXAMPLE #1

LEVELS OF MOLD SPORES (SPORES/CUBIC METER) FOUND INSIDE COMPARED TO ONE OUTSIDE CONTROL						
AREAS	Penicillium/ Aspergillus	Cladosporium	Basidiospores	Rare spores	Other spore types	TOTAL spores
OUTSIDE CONTROL	160	160	2,450	None	471	3,241
Dining room	160	213	533	None	107	1,013
Guest bathroom	53	107	160	None	40	360

RESULTS: *Mold spores levels found to be within the normal range for all genera. The elevated amount of Basidiospores found in the outside sample is normal. Basidiospores are related to mushrooms and their incidence spikes after a rainy day.*

EXAMPLE #2

LEVELS OF MOLD SPORES (SPORES/CUBIC METER) FOUND INSIDE COMPARED TO ONE OUTSIDE CONTROL						
AREAS	Penicillium/ Aspergillus	Cladosporium	Basidiospores	Rare spores	Other spore types	TOTAL spores
OUTSIDE CONTROL	320	320	3,520	None	359	4,519
Master bedroom	1,839	213	160	564 Stachybotrys	107	2,883
Guest bedroom	1,107	53	53	1,690 Stachybotrys & Chaetomium	40	2,943

RESULTS: *Serious mold problem found in both rooms with Stachybotrys and Chaetomium spores, even though the total amount of mold spores inside in each room is lower than the outside sample.*

EXAMPLE #3

LEVELS OF MOLD SPORES (SPORES/CUBIC METER) FOUND INSIDE COMPARED TO ONE OUTSIDE CONTROL

AREAS	Penicillium/ Aspergillus	Cladosporium	Basidiospores	Rare spores	Other spore types	TOTAL spores
OUTSIDE CONTROL	53	213	160	None	93	513
Family room, ambient air	14,940	213	160	None	107	15,420
Family room East wall cavity	73,000	53	125	530 Stachybotrys & Chaetomium	40	73,748
Family room North wall cavity	21,000	25	25	270 Chaetomium & Stachybotrys	40	21,360

RESULTS: *Mold problem in both walls. The elevated amount of spores (in red) found in the ambient room came from both walls.*

EXAMPLE #4

LEVELS OF MOLD SPORES (SPORES/CUBIC METER) FOUND INSIDE COMPARED TO ONE OUTSIDE CONTROL

AREAS	Penicillium/ Aspergillus	Cladosporium	Basidiospores	Rare spores	Other spore types	TOTAL spores
OUTSIDE CONTROL	107	213	107	None	693	1,120
Dining room	321	107	107	None	146	413
Office, ambient air	31,800	213	1,440	66 Chaetomium	292	33,811

RESULTS: *Mold problem in the office (in red). Elevated spores (in red). Wood rot seen on the ceiling beam and visible mold-like substance on the wall. Elevated amount of Basidiospores confirmed the wood rot.*

WHY IS STACHYBOTRYS DIFFERENT?

An air sample taken in this room revealed a normal amount of mold spores. The inspector noticed a small amount of mold-like substance and took a swab sample, which was identified by the laboratory as Stachybotrys. Remediation of the ceiling/wall was recommended. A large amount of Stachybotrys was found behind the wall and ceiling.

Having spoken about the importance of correctly assessing the air quality of a building through sampling, we are now going to tell you about *Stachybotrys,* the black slimy mold that is sometimes quite difficult to detect. It makes the job of mold inspectors tricky, and keeps them vigilant.

Stachybotrys grows in wet environments where food (organic material) is present. It is usually found when a source of water has been present for a long period of time. Like other molds, it grows and spreads by producing spores. However, it grows in a slimy mass and because of that the spores do not aerosolize easily. Unless the spores are airborne at the time of sampling, an air test may not pick up any of these spores. Therefore, it is possible to find no

Stachybotrys spores in an air sample while a source of mold is nearby, seen or unseen. If a few mold spores are found in an air sample and mold is not readily seen, the mold colony should be searched out in walls or other hidden spaces.

Stachybotrys produces toxins that are known to be toxic to humans and animals. Another mold often found with *Stachybotrys* is *Chaetomium* mold. Both can become a problem when they are found inside, even in low quantities. Depending on the location and the extent of contamination, the occupants may need to leave the house to avoid further exposure until the mold remediation has been completed. You could say that Stachy is on the black list (no pun intended) of every mold inspector.

Stachybotrys is responsible for bringing mold into the American consciousness. It all started with the Ballard family in Texas who won a multi-million dollar lawsuit against their insurance company. The jury found that the negligence of their insurance company caused them permanent health problems. Since then Mrs. Ballard has created a non-profit organization, "Policyholders of America" (POA), to keep homeowners informed about legislation issues and problems with insurance companies. www.policyholdersofamerica.org. See *INSURANCE TIPS.*

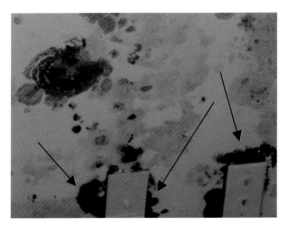

STACHYBOTRYS
(Black mold)

Just imagine the amount of mold present behind the switch plates in the wall cavity.

STACHYBOTRYS MOLD
CASE STUDIES

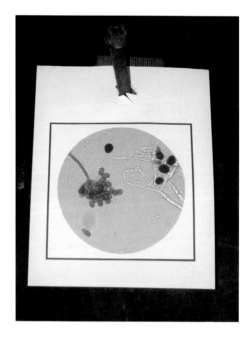

Stachybotrys is the genus name for what is commonly known as "black toxic mold." Its species include molds such as *Stachybotrys atra*, and *Stachybotrys chartarum*. Well known cases involving this type of mold were even featured on TV shows – 60 Minutes, and Forensic Files. Medical specialists were baffled as to why several people in a household were getting terribly sick. It turned out that the culprit was none other than *Stachybotrys* mold. Every mold inspector is on the lookout for this particular mold. It sometimes eludes even the most experienced inspector because it is often hidden from view and sometimes its presence cannot be detected by an air sample since it does not aerosolize easily.

The following scenarios will give you an idea on how easily this particular mold can be missed and how we can find a serious mold problem when we least expect it. We hope that these case studies will heighten your awareness of *Stachybotrys* mold.

For the sake of simplicity, we refer to "various types of mold" as XYZ, and we refer to *Stachybotrys* by name.

CASE SCENARIO #1

Mold found on a wall is identified by the laboratory as being XYZ and *Stachybotrys* mold. Mold remediation is recommended and the post-remediation testing indicates that the remediation has been successful. Case closed!

CASE SCENARIO #2

Observation: Water stain on a wall;

Recommendation: Collect an ambient air sample and a wall cavity sample with one outside control. Owner opted for one ambient air sample and one outside control to save money.

Sampling results: Elevated amount of XYZ found in the ambient air of the room.

Recommendation: Remediate the wall.

Post-remediation testing #1: Testing of the ambient air following remediation revealed a high level of *Stachybotrys* mold spores, even though *Stachybotrys* mold spores had not been found in the ambient air in the original test.

Conclusion: *Stachybotrys* mold was apparently behind the wall, and when the wall was remediated, *Stachybotrys* got jostled and thousands of spores went into the air.

Due to the elevated amount of spores, it is apparent that the mold remediation had not been carried out properly.

Recommendation: Additional mold remediation needed.

<u>Post-remediation testing #2</u>: It passed the clearance test, case closed.

In this case, we found out Stachybotrys mold purely by chance because no Stachy spores were found in the initial ambient air sample.

CASE SCENARIO #3

<u>Observation</u>: Occupants saw water coming in at the baseboard during a hurricane. The inspector saw a one-inch black spot at the top of the wall.

<u>Recommendation</u>: Collect an ambient air sample, a wall cavity sample, a surface sample, and an outside control.

<u>Sampling results</u>: The ambient air sample and wall cavity samples were both within the normal range for all types of mold, but the surface sample was identified as *Stachybotrys*.

<u>Recommendation</u>: Remediate the wall. Mold remediators found a high level of contamination inside the wall cavity, the insulation and the wood studs.

<u>Post-remediation testing</u>: Passed clearance test.

If the inspector had not noticed the one-inch black spot, there would have been nothing to swab and therefore the mold contamination would not have been found.

CASE SCENARIO #4

<u>Observation</u>: Client requested three air samples to test a building corridor – east and west ends and one in the middle.

Sampling results: All ambient air samples showed normal levels of XYZ, except for two single spores of *Stachybotrys* found in the west end sample.

Recommendation: Since we found a few spores of *Stachybotrys* we recommended a mold investigation of the west end. If a few spores of *Stachybotrys* were present, a mold colony had to be present.

Mold investigation: The initial walkthrough did not reveal anything out of the ordinary. We questioned the owner as to why he had wanted the sampling done in the first place. He pointed to his nose and said he had detected a slight odor at the east end of the hallway. We asked to go to the east end and started an inspection. At the back of a shower we found a small black spot on the baseboard. A pipe in the shower had leaked for a while and this was the source of *Stachybotrys*.

Recommendation: Remediate the shower wall.

Post-remediation testing: Passed clearance test.

Conclusion: No *Stachybotrys* spores were found in the east end sample where the actual problem was located. The two *Stachybotrys* spores found in the west end sample had somehow been transported through the corridor from the east end of the building. Even though two spores in almost any other types of mold would be inconsequential, finding a few spores of *Stachybotrys* indicated that a mold source likely existed somewhere in the building.

2. THE MOLD REMEDIATION PHASE

WHAT IS MOLD REMEDIATION?

Mold remediation, also called mold mitigation or abatement, involves the physical removal of the infested medium such as sheetrock, or scraping mold from wood studs and the removal of contaminated material <u>safely</u> from the premises. During this process mold will be disturbed and millions of spores will become airborne. It is thus essential that proper precautions be taken beforehand to protect the workers and the occupants, as well as preventing the mold spores from contaminating adjacent areas.

Believe it or not the basic principle of mold remediation has changed very little over two thousand years. See Cleansing From Mildew, Leviticus, Chapter 14:33-53 at the end of this book.

> *"He (the priest) must have all the inside walls of the house scraped and the material that is scraped off dumped into an unclean place outside the town."*

The Environmental Protection Agency and the New York City Department of Health Guidelines for Mold Remediation were pioneers in setting up standards for mold remediation. Since then, professional trade organizations have developed their own guidelines. Although there are slight variations between them mold remediation involves three main areas:

- Cleanup methods

- Personal Protective Equipment (PPE)

- Containment

MOLD REMEDIATION FLOWCHART

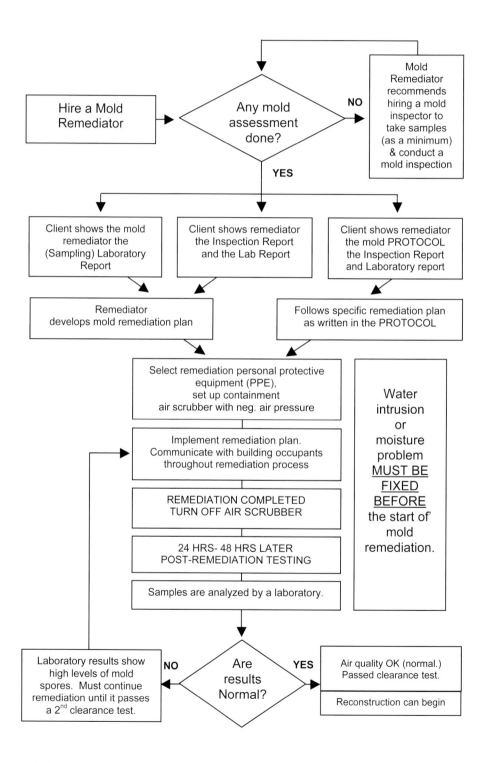

Hire a Mold Remediator

Any mold assessment done?

NO → Mold Remediator recommends hiring a mold inspector to take samples (as a minimum) & conduct a mold inspection

YES

Client shows the mold remediator the (Sampling) Laboratory Report

Client shows remediator the Inspection Report and the Lab Report

Client shows remediator the mold PROTOCOL the Inspection Report and Laboratory report

Remediator develops mold remediation plan

Follows specific remediation plan as written in the PROTOCOL

Select remediation personal protective equipment (PPE), set up containment air scrubber with neg. air pressure

Implement remediation plan. Communicate with building occupants throughout remediation process

REMEDIATION COMPLETED TURN OFF AIR SCRUBBER

24 HRS- 48 HRS LATER POST-REMEDIATION TESTING

Samples are analyzed by a laboratory.

Water intrusion or moisture problem <u>MUST BE FIXED BEFORE</u> the start of' mold remediation.

Are results Normal?

NO → Laboratory results show high levels of mold spores. Must continue remediation until it passes a 2nd clearance test.

YES → Air quality OK (normal.) Passed clearance test.

Reconstruction can begin

WHAT ARE THE QUALIFICATIONS
OF
A MOLD REMEDIATOR?

Mold remediation should not be taken lightly. It is very important to hire remediation professionals who are competent in removing mold safely. Hiring a "handyman" is not a good idea because mold is a serious matter and should be handled with the greatest care.

A distinction should be made between a mold inspector and a mold remediator. They perform two entirely different functions. Mold inspectors are more scientists/consultants, hired to assess the air quality in relation to mold, document the water damage and mold contamination and to recommend specific remediation. The remediator is a contractor, not a consultant, hired to clean mold from contaminated areas and prevent the spreading of spores to other parts of the building. In no instance should a remediator perform mold inspections or collect microbial samples. This can create a conflict of interest.

A mold remediator should have received training in mold remediation from an accredited school and passed a test that qualifies him as a mold remediator. In addition, he should be licensed, insured, and bonded, and preferably have experience.

- The mold remediation company should perform only mold remediation, not testing.

- The mold remediation company should be able to follow the protocol for mold remediation written by the mold investigation company, to remove the mold infested medium safely, and to dispose of the contaminated material.

- Determining remediation requirements for fungal infestation is based on the extent of visible contamination. It must be understood that the full assessment of the damage must be left to the professional judgment and practicality of the remediation company once abatement provides full visibility of the mold-infested medium. Further work or abatement may be deemed necessary at that time to successfully remove all contaminated materials.

- The mold remediation company may or may not repair water intrusion, such as leaks. In all situations, the underlying cause of water accumulation must be rectified or fungal growth will recur. Repairs to fix the water source must be done before mold remediation takes place. Some mold remediation companies have licensed professionals on their staff that can make the repairs, while others do not.

- Some mold remediation companies perform only mold remediation. Other remediation companies will also provide the "put back"; i.e. replace drywall, ceiling tiles, install cabinets, etc. and have professionals on staff who can do the work.

- A mold remediation company should stand behind its work. The crew should come back and perform more remediation in the event the final clearance test (post-remediation testing) fails the first try.

WHAT EQUIPMENT
IS USED IN MOLD REMEDIATION?

PERSONAL PROTECTIVE EQUIPMENT (PPE)

Mold remediators must wear protective clothing and gear when dealing with mold. Collectively these items are called Personal Protective Equipment.

CONTAINMENT

Containment is the process of isolating a mold-contaminated area with polyethylene sheeting to prevent spores from spreading to the rest of the home or building.

The containment area is set under "negative pressure" so that the mold spores floating in the air will be drawn inside the work area and be picked up and removed by the air scrubber. There can be full containment involving the entire room, or partial containment involving part of a room. The size and extent of mold contamination will dictate the type of containment.

NEGATIVE AIR PRESSURE

Negative air pulls outside air inward and prevents the contaminated air escaping the containment area. This is accomplished with either a negative air machine or a portable air scrubber. In order to create a pressure differential, at least 10% more volume of air is exhausted from the contaminated area compared to the volume of air supplied by the machine. Negative air machines are very bulky and heavy and are designed to HEPA filter large volumes of air on commercial sites while the portable air scrubber is used mainly in residential buildings.

PORTABLE AIR SCRUBBER

A portable air scrubber machine creates a negative air pressure and also cleans the air of mold spores with various size filters, including a HEPA filter.

Photo courtesy of Abatement Technologies, Inc.
All Rights Reserved.

NEGATIVE PRESSURE ILLUSTRATION

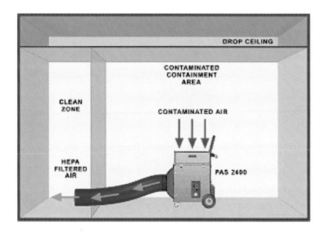

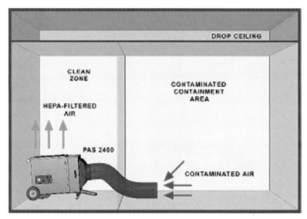

The air scrubber can be located inside the containment area or in an adjacent room if space is a problem or work area is at a remote location. In all instances the air is exhausted out of the containment area through flexible outlet ducting and out of the building.

SODA AND DRY ICE BLASTING

When mold has anchored itself in wood and its root-like structures (hyphae) are embedded in the material, mold must be sanded or scraped off, which is time consuming. On large areas, mold can be blasted out using high-power blasters thus saving time. Sodium bicarbonate and dry-ice are used as abrasives. Both processes are highly effective in mold removal while minimizing damage to the underlying surface. Blasting allows the operator to clean deep into small gaps and around corners. The operator may choose one method over the other depending on the job and on personal preference.

SODA BLASTING

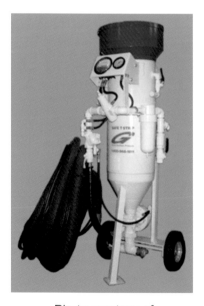

Photo courtesy of
Grand Northern Products. Inc.

BEFORE

AFTER

Photos courtesy of
First Alert Emergency Services

125

DRY ICE BLASTING

Dry ice blasting is also known as CO_2 Blasting.

Photo courtesy of
RSG Technologies, Inc.

Photo courtesy of RSG Technologies, Inc.

HEPA VACUUM

A HEPA vacuum is designed to remove airborne particles that settle onto floors and other surfaces.

LIGHWEIGHT HIP-PACK VACUUM

Variations of the HEPA vacuum

WHAT ARE THE EPA
MOLD REMEDIATION GUIDELINES?

The following mold remediation guidelines are reprinted from the publication titled: *MOLD REMEDIATION IN SCHOOLS AND COMMERCIAL BUILDINGS*. These guidelines do not apply solely for schools and commercial buildings; they apply to any type of mold remediation.

This free publication can be obtained by calling: 1-800-438-4318.

Various publications can also be found on the EPA WEBSITE: www.epa.gov/mold/moldresources.html

Table 1: Water Damage - Cleanup and Mold Prevention

Table 1 presents strategies to respond to water damage within 24-48 hours. These guidelines are designed to help avoid the need for remediation of mold growth by taking quick action before growth starts. If mold growth is found on the materials listed in Table 1, refer to Table 2 for guidance on remediation. Depending on the size of the area involved and resources available, professional assistance may be needed to dry an area quickly and thoroughly.

Please note that Tables 1 and 2 contain general guidelines. Their purpose is to provide basic information for remediation managers to first assess the extent of the damage and then to determine whether the remediation should be managed by in-house personnel or outside professionals. The remediation manager can then use the guidelines to help design a remediation plan or to assess a plan submitted by outside professionals.

Table 1: Water Damage - Cleanup and Mold Prevention

Guidelines for Response to Clean Water Damage within 24-48 Hours to Prevent Mold Growth*

Water-Damaged Material[†]	Actions
Books and papers	For non-valuable items, discard books and papers. Photocopy valuable/important items, discard originals. Freeze (in frost-free freezer or meat locker) or freeze-dry.
Carpet and backing - dry within 24-48 hours[§]	Remove water with water extraction vacuum. Reduce ambient humidity levels with dehumidifier. Accelerate drying process with fans.
Ceiling tiles	Discard and replace.
Cellulose insulation	Discard and replace.
Concrete or cinder block surfaces	Remove water with water extraction vacuum. Accelerate drying process with dehumidifiers, fans, and/or heaters.
Fiberglass insulation	Discard and replace
Hard surface, porous flooring[§] (Linoleum, ceramic tile, vinyl)	Vacuum or damp wipe with water and mild detergent and allow to dry; scrub if necessary. Check to make sure underflooring is dry; dry underflooring if necessary.
Non-porous, hard surfaces (Plastics, metals)	Vacuum or damp wipe with water and mild detergent and allow to dry; scrub if necessary.
Upholstered furniture	Remove water with water extraction vacuum. Accelerate drying process with dehumidifiers, fans, and/or heaters. May be difficult to completely dry within 48 hours. If the piece is valuable, you may wish to consult a restoration/water damage professional who specializes in furniture.
Wallboard	(Drywall and gypsum board) May be dried in place if there is no obvious swelling and the seams are intact. If not, remove, discard, and replace. Ventilate the wall cavity, if possible.
Window drapes	Follow laundering or cleaning instructions recommended by the manufacturer.
Wood surfaces	Remove moisture immediately and use dehumidifiers, gentle heat, and fans for drying. (Use caution when applying heat to hardwood floors.) Treated or finished wood surfaces may be cleaned with mild detergent and clean water and allowed to dry. Wet paneling should be pried away from wall for drying.

*If mold growth has occurred or materials have been wet for more than 48 hours, consult Table 2 guidelines. Even if materials are dried within 48 hours, mold growth may have occurred. Items may be tested by professionals if there is doubt. Note that mold growth will not always occur after 48 hours; this is only a guideline.

These guidelines are for damage caused by clean water. If you know or suspect that the water source is contaminated with sewage, or chemical or biological pollutants, then Personal Protective Equipment and containment are required by OSHA. An experienced professional should be consulted if you and/or your remediators do not have expertise remediating in contaminated water situations. Do not use fans before determining that the water is clean or sanitary.

[†] If a particular item(s) has high monetary or sentimental value, you may wish to consult a restoration/water damage specialist.

[§] The subfloor under the carpet or other flooring material must also be cleaned and dried. See the appropriate section of this table for recommended actions depending on the composition of the subfloor.

Table 2: Mold Remediation Guidelines

Table 2 presents remediation guidelines for building materials that have or are likely to have mold growth. The guidelines in Table 2 are designed to protect the health of occupants and cleanup personnel during remediation. These guidelines are based on the area and type of material affected by water damage and/or mold growth. Please note that these are guidelines; some professionals may prefer other cleaning methods.

If you are considering cleaning your ducts as part of your remediation plan, you should consult EPA's publication entitled, "Should You Have the Air Ducts In Your Home Cleaned?" If possible, remediation activities should be scheduled during off-hours when building occupants are less likely to be affected.

Although the level of personal protection suggested in these guidelines is based on the total surface area contaminated and the potential for remediator and/or occupant exposure, professional judgment should always play a part in remediation decisions. These remediation guidelines are based on the size of the affected area to make it easier for remediators to select appropriate techniques, not on the basis of health effects or research showing there is a specific method appropriate at a certain number of square feet. The guidelines have been designed to help construct a remediation plan. The remediation manager will then use professional judgment and experience to adapt the guidelines to particular situations. When in doubt, caution is advised. Consult an experienced mold remediator for more information.

In cases in which a particularly toxic mold species has been identified or is suspected, when extensive hidden mold is expected (such as behind vinyl wallpaper or in the HVAC system), when the chances of the mold becoming airborne are estimated to be high, or sensitive individuals (e.g., those with severe allergies or asthma) are present, a more cautious or conservative approach to remediation is indicated. Always make sure to protect remediators and building occupants from exposure to mold.

Table 2: Guidelines for Remediating Building Materials with Mold Growth Caused by Clean Water*

Material or Furnishing Affected	Cleanup Methods†	Personal Protective Equipment	Containment
SMALL - Total Surface Area Affected Less Than 10 square feet (ft2)			
Books and papers	3	Minimum N-95 respirator, gloves, and goggles	None required
Carpet and backing	1, 3		
Concrete or cinder block	1, 3		
Hard surface, porous flooring (linoleum, ceramic tile, vinyl)	1, 2, 3		
Non-porous, hard surfaces (plastics, metals)	1, 2, 3		
Upholstered furniture & drapes	1, 3		
Wallboard (drywall and gypsum board)	3		
Wood surfaces	1, 2, 3		
MEDIUM - Total Surface Area Affected Between 10 and 100 (ft2)			
Books and papers	3	Limited or Full Use professional judgment, consider potential for remediator exposure and size of contaminated area	Limited Use professional judgment, consider potential for remediator/occupant exposure and size of contaminated area
Carpet and backing	1,3,4		
Concrete or cinder block	1,3		
Hard surface, porous flooring (linoleum, ceramic tile, vinyl)	1,2,3		
Non-porous, hard surfaces (plastics, metals)	1,2,3		
Upholstered furniture & drapes	1,3,4		
Wallboard (drywall and gypsum board)	3,4		
Wood surfaces	1,2,3		
LARGE - Total Surface Area Affected Greater Than 100 (ft2) or Potential for Increased Occupant or Remediator Exposure During Remediation Estimated to be Significant			
Books and papers	3	Full Use professional judgment, consider potential for remediator/occupant exposure and size of contaminated area	Full Use professional judgment, consider potential for remediator exposure and size of contaminated area
Carpet and backing	1,3,4		
Concrete or cinder block	1,3		
Hard surface, porous flooring (linoleum, ceramic tile, vinyl)	1,2,3,4		
Non-porous, hard surfaces (plastics, metals)	1,2,3		
Upholstered furniture & drapes	1,3,4		
Wallboard (drywall and gypsum board)	3,4		
Wood surfaces	1,2,3,4		

132

Table 2	**Continued**

*Use professional judgment to determine prudent levels of Personal Protective Equipment and containment for each situation, particularly as the remediation site size increases and the potential for exposure and health effects rises. Assess the need for increased Personal Protective Equipment, if, during the remediation, more extensive contamination is encountered than was expected. Consult Table 1 if materials have been wet for less than 48 hours, and mold growth is not apparent. These guidelines are for damage caused by clean water. If you know or suspect that the water source is contaminated with sewage, or chemical or biological pollutants, then the Occupational Safety and Health Administration (OSHA) requires PPE and containment. An experienced professional should be consulted if you and/or your remediators do not have expertise in remediating contaminated water situations.

†Select method most appropriate to situation. Since molds gradually destroy the things they grow on, if mold growth is not addressed promptly, some items may be damaged such that cleaning will not restore their original appearance. If mold growth is heavy and items are valuable or important, you may wish to consult a restoration/water damage/remediation expert. Please note that these are guidelines; other cleaning methods may be preferred by some professionals.

Cleanup Methods

***Method 1:** Wet vacuum (in the case of porous materials, some mold spores/fragments will remain in the material but will not grow if the material is completely dried). Steam cleaning may be an alternative for carpets and some upholstered furniture.

***Method 2:** Damp-wipe surfaces with plain water or with water and detergent solution (except wood —use wood floor cleaner); scrub as needed.

***Method 3:** High-efficiency particulate air (HEPA) vacuum after the material has been thoroughly dried. Dispose of the contents of the HEPA vacuum in well-sealed plastic bags.

***Method 4:** Discard _ remove water-damaged materials and seal in plastic bags while inside of containment, if present. Dispose of as normal waste. HEPA vacuum area after it is dried.

Personal Protective Equipment (PPE)

Minimum: Gloves, N-95 respirator, goggles/eye protection
*Limited: Gloves, N-95 respirator or half-face respirator with HEPA filter, disposable overalls, goggles/eye protection
*Full: Gloves, disposable full body clothing, head gear, foot coverings, full-face respirator with HEPA filter

Containment

Limited: Use polyethylene sheeting ceiling to floor around affected area with a slit entry and covering flap; maintain area under negative pressure with HEPA filtered fan unit. Block supply and return air vents within containment area.
*Full: Use two layers of fire-retardant polyethylene sheeting with one airlock chamber. Maintain area under negative pressure with HEPA filtered fan exhausted outside of building. Block supply and return air vents within containment area.

Table developed from Literature and remediation documents including Bioaerosols: Assessment and Control (American Conference of Governmental Industrial Hygienists, 1999) and IICRC S500, Standard and Reference Guide for Professional Water Damage Restoration, (Institute of Inspection, Cleaning and Restoration, 1999.)

Cleanup Methods

A variety of mold cleanup methods are available for remediating damage to building materials and furnishings caused by moisture control problems and mold growth. The specific method or group of methods used will depend on the type of material affected, as presented in Table 2. Please note that professional remediators may use some methods not covered in these guidelines; absence of a method in the guidelines does not necessarily mean that it is not useful.

Method 1: Wet Vacuum

Wet vacuums are vacuum cleaners designed to collect water. They can be used to remove water from floors, carpets, and hard surfaces where water has accumulated. They should not be used to vacuum porous materials, such as gypsum board. They should be used only when materials are still wet—wet vacuums may spread spores if sufficient liquid is not present. The tanks, hoses, and attachments of these vacuums should be thoroughly cleaned and dried after use since mold and mold spores may stick to the surfaces.

Method 2: Damp Wipe

Whether dead or alive, mold is allergenic, and some molds may be toxic. Mold can generally be removed from nonporous (hard) surfaces by wiping or scrubbing with water, or water and detergent. It is important to dry these surfaces quickly and thoroughly to discourage further mold growth. Instructions for cleaning surfaces, as listed on product labels, should always be read and followed. Porous materials that are wet and have mold growing on them may have to be discarded. Since molds will infiltrate porous substances and grow on or fill in empty spaces or crevices, the mold can be difficult or impossible to remove completely.

Method 3: HEPA Vacuum

HEPA (High-Efficiency Particulate Air) vacuums are recommended for final cleanup of remediation areas after materials have been thoroughly dried and contaminated materials removed. HEPA vacuums are also recommended for cleanup of dust that may have settled on surfaces outside the remediation area. Care must be taken to assure that the filter is properly seated in the vacuum so that all the air must pass through the filter. When changing the vacuum filter, remediators should wear PPE to prevent exposure to the mold that has been captured. The filter and contents of the HEPA vacuum must be disposed of in well-sealed plastic bags.

Method 4: Discard — Remove Damaged Materials and Seal in Plastic Bags

Building materials and furnishings that are contaminated with mold growth and are not salvageable should be double-bagged using 6-mil polyethylene sheeting. These materials can then usually be discarded as ordinary construction waste. It is important to package mold-contaminated materials in sealed bags before removal from the containment area to minimize the dispersion of mold spores throughout the building. Large items that have heavy mold growth should be covered with polyethylene sheeting and sealed with duct tape before they are removed from the containment area.

Personal Protective Equipment (PPE)

If the remediation job disturbs mold and mold spores become airborne, then the risk of respiratory exposure goes up. Actions that are likely to stir up mold include: breakup of moldy porous materials such as wallboard; invasive procedures used to examine or remediate mold growth in a wall cavity; actively stripping or peeling wallpaper to remove it; and using fans to dry items.

The following sections discuss the different types The primary function of Personal Protective Equipment (PPE) is to avoid inhaling mold and mold spores and to avoid mold contact with the skin or eyes. Please note that all individuals using certain PPE equipment, such as half-face or full-face respirators, must be trained, must have medical clearance, and must be fit-tested by a trained professional. In addition, the use of respirators must follow a complete respiratory protection program as specified by the Occupational Safety and Health Administration.

Skin and Eye Protection
Gloves are required to protect the skin from contact with mold allergens (and in some cases mold toxins) and from potentially irritating cleaning solutions. Long gloves that extend to the middle of the forearm are recommended. The glove material should be selected based on the type of materials being handled. If you are using a biocide (such as chlorine bleach) or a strong cleaning solution, you should select gloves made from natural rubber, neoprene, nitrile, polyurethane, or PVC. If you are using a mild detergent or plain water, ordinary household rubber gloves may be used. To protect your eyes, use properly fitted goggles or a full-face respirator with HEPA filter. Goggles must be designed to prevent the entry of dust and small particles. Safety glasses or goggles with open vent holes are not acceptable.

Respiratory Protection
Respirators protect cleanup workers from inhaling airborne mold, mold spores, and dust.

Minimum: When cleaning up a small area affected by mold, you should use an N-95 respirator. This device covers the nose and mouth, will filter out 95% of the particulates in the air, and is available in most hardware stores. In situations where a full-face respirator is in use, additional eye protection is not required.

Limited: Limited PPE includes use of a half-face or full-face air purifying respirator (APR) equipped with a HEPA filter cartridge. These respirators contain both inhalation and exhalation valves that filter the air and ensure that it is free of mold particles. Note that half-face APRs do not provide eye protection. In addition, the HEPA filters do not remove vapors or gases. You should always use respirators approved by the National Institute for Occupational Safety and Health.

Full: In situations in which high levels of airborne dust or mold spores are likely or when intense or long-term exposures are expected (e.g., the cleanup of large areas of contamination), a full-face, powered air purifying respirator (PAPR) is recommended. Full-face PAPRs use a blower to force air through a HEPA filter. The HEPA-filtered air is supplied to a mask that covers the entire face or a hood that covers the entire head. The positive pressure within the hood prevents unfiltered air from entering through penetrations or gaps. Individuals must be trained to use their respirators before they begin remediation. The use of these respirators must be in compliance with OSHA regulations.

Disposable Protective Clothing
Disposable clothing is recommended during a medium or large remediation project to prevent the transfer and spread of mold to clothing and to eliminate skin contact with mold.

Limited: Disposable paper overalls can be used.

Full: Mold-impervious disposable head and foot coverings, and a body suit made of a breathable material, such as TYVEK®, should be used. All gaps, such as those around ankles and wrists, should be sealed (many remediators use duct tape to seal clothing).

Containment

The purpose of containment during remediation activities is to limit release of mold into the air and surroundings, in order to minimize the exposure of remediators and building occupants to mold. Mold and moldy debris should not be allowed to spread to areas in the building beyond the contaminated site.

The two types of containment recommended in Table 2 are limited and full. The larger the area of moldy material, the greater the possibility of human exposure and the greater the need for containment. In general, the size of the area helps determine the level of containment. However, a heavy growth of mold in a relatively small area could release more spores than a lighter growth of mold in a relatively large area. Choice of containment should be based on professional judgment. The primary object of containment should be to prevent occupant and remediator exposure to mold.

Limited Containment

Limited containment is generally recommended for areas involving between 10 and 100 square feet (ft^2) of mold contamination. The enclosure around the moldy area should consist of a single layer of 6-mil, fire-retardant polyethylene sheeting. The containment should have a slit entry and covering flap on the outside of the containment area. For small areas, the polyethylene sheeting can be affixed to floors and ceilings with duct tape. For larger areas, a steel or wooden stud frame can be erected and polyethylene sheeting attached to it. All supply and air vents, doors, chases, and risers within the containment area must be sealed with polyethylene sheeting to minimize the migration of contaminants to other parts of the building. Heavy mold growth on ceiling tiles may impact HVAC systems if the space above the ceiling is used as a return air plenum. In this case, containment should be installed from the floor to the ceiling deck, and the filters in the air handling units serving the affected area may have to be replaced once remediation is finished.

The containment area must be maintained under negative pressure relative to surrounding areas. This will ensure that contaminated air does not flow into adjacent areas. This can be done with a HEPA-filtered fan unit exhausted outside of the building. For small, easily contained areas, an exhaust fan ducted to the outdoors can also be used. The surfaces of all objects removed from the containment area should be remediated/cleaned prior to removal. The remediation guidelines outlined in Table 2 can be implemented when the containment is completely sealed and is under negative pressure relative to the surrounding area.

Full Containment

Always maintain the containment area under negative pressure. Exhaust fans to outdoors and ensure that adequate makeup air is provided. If the containment is working, the polyethylene sheeting should billow inwards on all surfaces. If it flutters or billows outward, containment has been lost, and you should find and correct the problem before continuing your remediation activities.	**Photo 8:** **Full containment area** **On large job** 

Full containment is recommended for the cleanup of mold-contaminated surface areas greater than 100 ft^2 or in any situation in which it appears likely that the occupant space would be further contaminated without full containment. Double layers of polyethylene should be used to create a barrier between the moldy area and other parts of the building. A decontamination chamber or airlock should be constructed for entry into and exit from the remediation area. The entryways to the airlock from the outside and from the airlock to the main containment area should consist of a slit entry with covering flaps on the outside surface of each slit entry. The chamber should be large enough to hold a waste container and allow a person to put on and remove PPE. All contaminated PPE, except respirators, should be placed in a sealed bag while in this chamber. Respirators should be worn until remediators are outside the decontamination chamber. PPE must be worn throughout the final stages of HEPA vacuuming and damp-wiping of the contained area. PPE must also be worn during HEPA vacuum filter changes or cleanup of the HEPA vacuum.

2. THE POST-REMEDIATION PHASE

WHAT IS A CLEARANCE TEST?
(POST REMEDIATION SAMPLING)

Post-remediation testing, also known as a clearance test, insures that the mold remediation has been performed properly, as evidenced by laboratory results. After remediation is completed, but before containment is removed and before any reconstruction takes place, i.e. installing new drywall, post remediation testing is positively, absolutely, unequivocally necessary. Period. No matter how "clean" the area may look, skipping post-remediation testing is not an option.

Post-remediation sampling should be done no earlier than 24 hours and preferably 48 hours after remediation is completed, during which time the windows and doors must remain closed and any air cleaning devices, i.e. air cleaners/purifiers/scrubbers, must be turned off. Understand that if any one of these conditions is not met, the test results could be adversely affected.

Air testing is conducted in areas that were remediated. The testing is carried out inside the containment area while the containment is still intact, and at least one air sample, preferably two, is/are compared to one outside sample. The post-remediation sampling protocol is the same as in the original sampling. It is crucial that the clearance test be performed by the same company that carried out the original testing. This will ensure that the same methodology is followed - the same volume of air collected, the same brand of spore traps employed, and the services of the same laboratory utilized. This will provide greater accuracy when comparing the levels of mold spores before and after remediation.

If the laboratory results show an elevated amount of mold spores, further remediation will be necessary. An elevated amount of spores in post-remediation testing can mean one of three things:

1. The mold remediators were not well qualified and did the job poorly.

2. Competent mold remediators have either missed something, or did not remediate enough. That happens.

3. Not enough air filtration (air scrubber operation).

WHO WILL PAY FOR A SECOND POST-REMEDIATION TESTING?

Hopefully you have this contingency written in your contract. If you don't, talk to the mold remediators <u>before</u> (the first) post-remediation testing takes place, and tell them you expect their work to pass the first time and to let you know when they are ready for testing. Talk to them about the eventuality that their work does not pass the first time. After all, if it does not pass you should not have to pay for a second testing. If you take this attitude, the remediators will make doubly sure that the first post-remediation test will pass with flying colors. Regardless, it is always best to establish those contingencies beforehand to avoid any disputes or ill feelings.

In the event the post-remediation samples (clearance test) do not pass the first time, the remediators will need to do more work. This additional work should not cost you any more, unless they find something major to remediate that they had not taken into account in their original bidding. Otherwise, this is part of their work. They will let you know when they are ready for a second testing.

After obtaining air clearance, showing that the mold spores inside the containment is within normal range for all genera compared to one outside sample, the containment can be dismantled and reconstruction can begin.

DEAD SPORES AND LIVE SPORES

Often, wannabe remediators believe, that because they have removed moldy drywall, the remediation will pass a clearance test. After all, nobody sees anything so therefore it must be clean. What they fail to realize is that the process of sampling involves collecting microscopic spores from the air that are invisible to the naked eye. So, unless the remediators have used an air scrubber to clean the air by removing the mold spores as they worked, the spores will remain in the air. We call these "residual" mold spores and the remediators must continue to run the air scrubbers for several days after the remediation is completed.

If something is dead it cannot hurt you, right? Wrong! Dead (non-viable) spores can affect people's health just as much as "live" (viable) spores, because we inhale them as we breathe and they can both produce allergenic type reactions. Once we were called in for post-remediation testing following a "miracle cure" used by a wannabe mold remediator. He had purchased a small fumigation apparatus and "fumigated" a house to "kill" all the mold spores. We were anxious to find out the results, as we like to keep an open mind. Who knows - someone may have found the miracle cure. For us, what counts are the laboratory results. Unfortunately, the laboratory results revealed a high level of mold spores was still present in the house. The so-called "mold remediator" got mad at us because his product says that it should kill all mold spores. He wanted to know what number of spores were dead and what number were alive. We told him that there was no way to separate the dead ones from the live ones.

As of the writing of this book, traditional mold remediation, which is the actual removal of contaminated materials, is the preferred choice among professional mold remediators.

INSURANCE TIPS

INSURANCE TIPS INVOLVING
WATER DAMAGE

There are lessons to be learned from past natural disasters. As a homeowner, there are several things you can do to minimize the damage caused by water intrusion. Water can cause tremendous damage and mold settling in on wet organic materials only compounds the problem. Considering that mold can start growing within 24 to 48 hours following water intrusion, it is important to take action immediately and to document everything. See *WHAT TO DO WITH WET BUILDING MATERIALS AND FURNISHINGS.*

Hurricanes and floods bring much misery and frustration to real estate owners who are desperate for anyone to come and assess the damage. Insurance adjusters seem slow to respond but this is because of the large number of claims. Many home and building owners may feel the need to preserve evidence and leave things as they are until an insurance adjuster comes to inspect the premises. In the meantime, nature takes its course and mold starts growing on building materials. In some cases, respiratory distress cause occupants to leave their homes.

Some homeowners call drying companies. Unfortunately, those companies are also overwhelmed with work, and they may not get on site right away. In many cases the delay will allow mold to grow. This brings up another dilemma: if mold is present and containment is lacking, the powerful fans used for drying could spread mold spores from contaminated materials to other parts of the home.

Taking things in your own hands during the first 24 hours has never been so critical since recently many insurance companies are excluding mold damage while continuing to cover water damage. This exclusion does not make sense because mold can grow only when moisture is present. This action will undoubtedly cause much grief to all concerned. Mold contaminated materials will be handled by handymen, not by trained mold remediators. Lack of personal protective gear will expose workers and occupants to potentially toxic mold and possibly make them sick. Lack of proper equipment and lack of containment during the demolition will cause millions of mold spores to be released into the air and contaminate the rest of the house. Lawyers will be busy.

We advise you to read your insurance policy now. Do not wait until you need to submit a claim to find out what is included or excluded in your policy. You may want to add mold coverage as an option. Become familiar with the terms of your insurance policy and extent of coverage. If you feel you have insufficient protection, contact your insurance company and make the necessary changes NOW, before disaster strikes.

Do your homework and follow these tips:

- Assess the damage and report it to your insurance company immediately.

- Keep track of whom you talk to with names, telephone numbers, and dates.

- Take as many pictures of the damage as you can and date the pictures.

- If possible, obtain an infrared survey. This will document, usually with pictures or video, where water intrusion took place. Keep in mind that IR surveys are useful only when the material is currently wet. Over time the material will dry out and water intrusion will no longer be visible in infrared.

- As time goes on, take more pictures to document the deterioration of material, water stains, or what may appear to be visible mold. Date the pictures.

- Keep all your receipts for expenses, such as tarps, etc.

INSURANCE TIPS INVOLVING MOLD

If time has elapsed and mold has had a chance to grow, you must be prepared to provide your insurance company with proof of mold contamination and damage. Many insurance companies will not pay or reimburse homeowners for the initial samples. This is easy to understand. As long as you do not produce a report that says there is mold growing inside the residence, they will assume there is no mold. If there is no mold, there is nothing to fix, and therefore nothing to pay. First, try to get authorization from your insurance company to hire a mold inspector. If they give you a hard time, it would behoove you to hire one at your own cost to collect samples and establish if there is a mold problem. Should a problem be found, your insurance company will not be able to contest the findings of a scientific laboratory report.

Mold can make people sick based upon their own sensitivity, the amounts and species of mold present. If you feel you or a member of your family is getting sick because of mold, move out immediately.

We recommend that mold samples be taken in places where visible mold is present as well as taking air samples in rooms where mold is suspected. If you are on a tight budget, we recommend taking an air sample in the worst room of the house with one outside control. After the laboratory report comes back confirming an elevated amount of mold spores, the insurance company will be more likely to pay for the cost of additional samples recommended by the mold inspector.

Many insurance companies will also want to see a mold inspection report that documents signs of water intrusion and visible mold. Some insurance companies will go a step further and request a mold remediation protocol. See *WHAT IS A MOLD REMEDIATION PROTOCOL?*

If you have any disagreement with your insurance company, write a letter addressed to your adjuster and if possible, include pictures. Don't forget to send the letter certified with return receipt.

If you have problems with your insurance company with regards to your claims, you can contact "Policy holders of America" (POA), www.policyholdersofamerica.org. This is a great place where you can get self-help with insurance claims and a wealth of useful information.

Mrs. Ballard, President of POA, advises:
>While most insurance policies exclude or cap mold remediation costs, most do not exclude the cost to repair water damage. Mold is always a consequence of water damage. Policyholders of America can help homeowners get paid for repairs when water damage occurs even when the insurer balks at coverage. Knowing how to file and document a claim is critical and Policyholders of America can guide homeowners through the maze for free.

It is also a good idea to take inventory of your possessions BEFORE a disaster strikes. See *PREVENTION – INVENTORY YOUR POSSESSIONS*.

REAL ESTATE TRANSACTION TIPS

GENERAL TRANSACTION TIPS

We constantly reassure sellers and buyers that a mold problem can always be fixed. If the buyer likes the house, discovering a mold problem is usually not a good reason to pass it up. The seller should have the leak repaired, the mold problem remediated by professional mold remediators, and with a successful clearance test the house can be sold.

Many people would not dream of buying a home without having a home inspection. The home inspector inspects systems such as the electrical system, plumbing, and HVAC (heat and A/C). His inspection insures that everything is in working order and up to code.

Not too many people have grasped the consequences of finding mold in a home. It is similar in scope to finding termites – fixable but depending on the extent of the damage, it can be costly.

Sometimes sellers have no idea that they have a mold problem and they are eager to get things repaired once a mold issue reveals itself. If they possess mold insurance, the mold remediation and repairs will be covered up to their coverage limit. Sometimes, the home is sold as-is. In this case negotiation may satisfy both parties.

Mold problems can happen anytime, anywhere, and to anybody's house. Water intrusion or leaks can happen, which may not be discovered right away. The key is to fix problems as soon as they are discovered and to practice prevention.

Mold sampling will benefit both the seller and the buyer. As a buyer it is always better to know if a home has a mold problem so that it can be fixed prior to closing. As a seller, you have the peace of mind that at the time of closing the air quality of your home was normal in relation to mold.

SELLER BEWARE

If in your state there is a mold clause in your real estate contract, answer truthfully about prior leaks. Everyone has had leaks at one time or another. If you answer "no", it may look suspicious. Just be truthful - your prospective buyer will appreciate your honesty.

If the buyer wants a mold inspection or have the home sampled to evaluate the air quality, do not panic. Instead, welcome the opportunity. If you have been vigilant about correcting problems chances are your home will pass with flying colors. If a problem exists, it is better that it is found <u>prior</u> to closing so that you can get it remediated, along with repairing the source of moisture. Following mold remediation, get post-remediation testing performed. With a clearance test in hand, no one can come back later and say that there was mold in the house when they bought it.

As a selling point, prior to putting your home on the market have your home tested by a professional mold inspector. If your house passes the first time, put your house on the market and advertise the fact that you have a mold report showing normal air quality. It is a great selling point.

If a problem is found, get it fixed <u>prior</u> to putting your home on the market. This is better than loosing a prospective buyer to delays inherent in remediation. After remediation has been completed, put your house on the market. Now you can show your prospective buyers the original testing, the scope of work performed, and finally you show the clearance test. This will show your buyers that you care and you did the right thing.

BUYER BEWARE

If there is a mold provision in your real estate contract, go one step further - ask the owner when they last had the roof replaced, which rooms were recently repainted, any prior leaks – bathrooms, kitchen, A/C, etc. Ask as many questions as you can. Ask if the stucco is EIFS (Exterior Insulating Finishing System), which could be potentially bad news if it lacks a drainage plane. If it is, get it inspected by a building forensic professional to make sure it is ok.

When you hire a home or building inspector, ask him to thoroughly check the walls for moisture. Ask him to note any of the red flags mentioned in this book (See *RED FLAGS*). Once the home/building inspection is completed, review the report thoroughly. Hire a reputable mold inspector to collect air samples in the questionable areas noted in the inspection report. If nothing was noted in the report hire a mold inspector to collect several samples throughout the house or building to test the air quality with respect to mold.

Experience has shown us that 99% of properties on the market look pristine (regardless on how they looked a few months prior). The smell of new paint usually permeates the air, and often the scent of burning candles or freshener makes people feel right at home. Within a few weeks the new owners start noticing a musty smell and start developing allergies. Within a month or two they'll start seeing a discoloration on the walls, and finally they will see mold coming through. You guessed it - the sellers repainted over moldy walls.

Most of the time the building will appear to be in perfect condition when it is put on the market. No matter how clean it may look, get samples taken and analyzed. We have found elevated amounts of mold spores in million dollar homes that looked spotless. A mold inspection consists mainly of a visual inspection and testing the walls for moisture. We will see what you see. The lack of moisture in walls is not always reliable since mold may be present in a wall that was once wet and is now dry. However, by collecting and analyzing several air samples throughout the building you will know scientifically what the occupants are breathing. If you can also afford an inspection, get both – inspection and samples. Purchasing a home is a big investment; get the home tested for mold. You will not regret it.

PREVENTION

MOLD PREVENTION

HOW TO DETECT LEAKS WITH
WATER ALARMS

Mold prevention is all about water control. Homeowners are familiar with a smoke alarms and most people wouldn't dream of doing without them in their homes. Few people, however, know that leak detectors even exist. Similar to a smoke detector that sounds an alarm when smoke is present, leak detectors sound an alarm when water is present. All homes will have leaks at one time or another. The question is not if, but when they will occur.

There are two main types of leak detector systems that alert the occupants – active and passive.

ACTIVE

The active detector is a system that is wired throughout the house. It not only beeps at a central location when a leak occurs, it turns off the water automatically. Some systems use moisture sensors, others a flow sensor coupled with a timer to activate the shut-off valve. Some systems can be connected to a home security system and programmed to call a monitoring company.

Such systems are very desirable, especially if a leak occurs when you are away from home.. This type of system is ideal but it is rather expensive to have the system wired and installed. Nevertheless it is a good investment. More information can be found on the Internet.

PASSIVE

The passive detector is a system made up of individual, self-contained units that are battery-operated and are placed in areas prone to leaks. This system is very affordable, and for the cost of a few pizzas, you can protect your entire home from leaks. Once the units are in place, you simply change the batteries according to instructions. Some will beep to warn you that the batteries are low and need to be replaced.

There are various brands that operate in the same fashion. The units are placed throughout the home next to appliances and plumbing connections. An alarm will sound upon contact with water. We recommend five units for an apartment and ten units for an average home. This system is simple and affordable. It is best to choose small units – some of them are bulky and impractical in small spaces. Follow the manufacturer's directions.

CONCEPT OF A LEAK DETECTOR UNIT

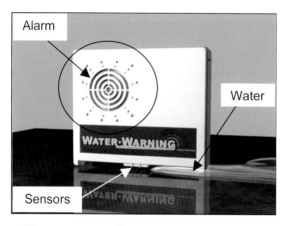

Alarm

Water

Sensors

1. Water appears

2. Contact is made at the sensors

3. Alarm sounds

Photo courtesy of www.waterwarning.net

The following appliances are prone to leaks:

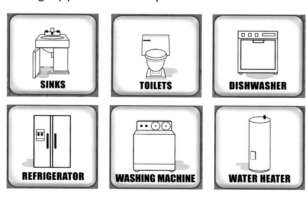

SINKS TOILETS DISHWASHER

REFRIGERATOR WASHING MACHINE WATER HEATER

Suggested placement of the units:

▪ Under the kitchen sink ▪ Under bathroom sinks ▪ Inside shower walls (as explained on next page) ▪ Behind toilets ▪ Underneath the dishwasher ▪ Next to the water softener	▪ Behind the refrigerator ▪ Behind the washing machine ▪ Next to the water heater ▪ Below the air handler (A/C) condensate pan ▪ In the basement ▪ Water filtration

We mentioned earlier in the section *WHAT ARE THE CONDITIONS FOR MOLD GROWTH?* how builders hide pipes from view. They do that for a good reason - pipes are ugly. However, the building industry should really look at ways homeowners could perform plumbing maintenance checks because every pipe has the potential for leaks. Because they are hidden from view, leaks may not be detected for months. This means that along with water damage to building materials, mold can settle in and contaminate surrounding areas. We find that if a slow leak goes undetected for a long period of time, such as a shower plumbing leak, toxic mold like *Stachybotrys* will likely become the predominant type of mold. Now, your health is at stake.

After seeing so many problems with shower leaks, including our own (oh yes, we had leaks too) we came up with a simple and inexpensive solution – have a small piece of drywall held by screws, as shown on next page, or a small wooden door, installed behind the shower or next to the bathtub. We placed two water-warning units at the base of the plumbing - one on each side of pipes because our floor is not even. Two units may be overkill, but having lived through a nightmare once, we want to make doubly sure that it will not happen again.

Sheetrock covers the opening in our closet wall.
This allows access to the wall cavity of our shower and lets us
replace the batteries of our water detectors.

Repairing a leak is cheap. The cost of fixing damage caused by water and mold is another matter. This can run into tens of thousands of dollars depending on the situation. Slow leaks are usually not covered by home insurance. We are not trying to scare you, but we want to impress upon you that you must take matters into your own hands and practice prevention. A leak detector system, whether active or passive is a must in every home. Cost cannot be cited as an excuse since individual units are inexpensive.

Homeowners having rental properties should also get water alarm detectors for their rental homes or apartments. Do not count on your tenants to change the batteries. Do it yourself.

Once your water alarm system is in place, check with your insurance company for possible premium discounts.

PS – Remember us in your prayers after you have followed our recommendations and you hear one of your units beep.

TIPS ON CONTROLLING
MOISTURE AND WATER INTRUSION

Mold prevention is about controlling moisture and water intrusion, and making timely repairs. It therefore behooves the home or building owner to perform simple periodic maintenance checks to prevent or catch leaks early and minimize the chances for mold growth. Responding quickly is of the utmost importance.

As you read the preventive measures outlined in this chapter, keep in mind the locations where a water alarm unit (discussed in the previous chapter) would be helpful. Suggestions regarding their locations are indicated with the acronym "WW" (water warning) to mean a device that will warn you of a leak. The following list of recommendations is not exhaustive.

INSIDE YOUR HOME

BATHROOMS
UNDER THE SINKS AND BEHIND TOILETS
Check under the sinks of your bathrooms for any possible leaks. Check the plumbing behind toilets. Repair immediately if defective. (WW)

SHOWER HEADS
Every few years have a plumber replace the shower arm in your showers. With time the shower arm can corrode, and leaks can start behind walls without your knowing it. Shower plumbing is notorious for leaks and the damage caused is tremendous. (WW)

SQUEEGEE AND SPONGE
To keep the humidity down, it is a good idea to use a squeegee to remove the excess water from the shower walls and shower door prior to stepping out

of the shower. A big sponge finishes the job. This takes only a minute.

CHECK GROUT and CAULKING
Check the grout of your shower(s) regularly and replace it when it begins to become thin. Do not wait to see holes in order to make repairs. Similarly, check the caulking around bathtubs regularly.

VENTILATION
Did you know that showering releases approximately one cup of water into the air? If you have a window in the bathroom, open it while taking a shower to let steam escaped, or use a ceiling exhaust fan, which should vent to the outside, and not into the attic.

FUNGICIDE
It is a good idea to include fungicide in the bathroom paint to discourage mold from growing on walls and ceilings. Ask your paint/hardware man.

WALLPAPER
Vinyl wallpapers, which are impermeable, and "washable" can lead to mold problems, whether in the bathroom or anywhere else in the house.

Do not use wallpapers (permeable or impermeable) in the bathroom. With time they will peel off because of high humidity and mold will grow on the back of the wallpaper.

CLOSETS
POOR VENTILATION
Poor ventilation in closet can be remedied by leaving the door open or installing a fan.

LOW TEMPERATURE
In cold climates, closets can become very cold, which can lead to mold growth (if RH is high). Leaving a 150-watt bulb on can help warm up the

closet. Closets can become crowded - so make sure that no piece of clothing or box can touch or fall on the bulb.

HUMIDITY FROM THE SLAB
People tend to stack boxes and personal effects on the perimeter of the closet. The humidity from the slab will be transferred to the boxes and personal effects. To remedy this, put your personal effects on small shelves or racks elevated off the floor.

KITCHEN
SINK
Keep the area under the sink organized. Clutter can hide small leaks for a long period of time and a serious mold problem can develop. (WW)

DISHWASHER
From time to time, unscrew the bottom front cover and look under the dishwasher with a flashlight for leaks. (WW)

REFRIGERATOR
If your refrigerator is equipped with an icemaker and water dispenser it is essential that you pull the refrigerator out from time to time to make sure that no leaks are developing. (WW)

HEATING, VENTILATION, AIR CONDITIONING (HVAC)
AIR FILTERS
A dirty HVAC system will not necessarily promote mold growth unless there is a source of water or elevated amount of humidity present. Dust is food for mold. Keeping an HVAC system clean is essential for clean air. We recommend obtaining a pleated paper filter with a MERV rating of 8. The MERV (Minimum Efficiency Rating Value) rating was implemented by the American Society of Heating, Refrigerating and Air-Conditioning Engineers (ASHRAE) to provide a standard for the size of the

pores in filter systems. A rating of 8 is small enough to aid in catching particles, dust, etc. while remaining large enough to permit good airflow.

A/C MAINTENANCE

We cannot overstress the importance of getting the HVAC system checked and serviced regularly, as it is a possible reservoir for mold growth. An HVAC system in poor working condition can cause mold to grow because of high humidity in the air, and, in turn, can spread that problem throughout the house.

RELATIVE HUMIDITY

Ask your A/C professional to check the relative humidity inside your home. The relative humidity should be kept below 60%. If the reading is continuously above 60%, it may be time to change your system, or get a dehumidifier to supplement your A/C.

DEHUMIDIFIER

Some molds do not need an active leak in order to grow. Some can take the humidity from the air and start growing on any surface. The relative humidity of the air should always be kept below 60%. An older A/C unit may not operate efficiently in removing the humidity from the air. When this happens it is time to replace the unit. If you are unable to purchase a new A/C for budgetary reasons, a stand alone dehumidifier will serve you well until such time you are able to do so.

NORTHERNERS' MISTAKE

Many Northerners who purchase property in sub-tropical climates make the mistake of leaving their A/C off when they return up North. When they come back six months or a year later, they find their vacation home inundated with mold. This is because some molds can simply take the moisture they need from the high humidity in the air.

OVERSIZED A/C
An oversized A/C will cool a house quickly and, in so doing, it will not remove enough moisture from the air.

MULTIPLE THERMOSTATS
The problem can be compounded for larger homes that possess multiple air handlers. Not only is the probability for mold growth increased, but the placement of the thermostats can also create problems of their own. If two thermostats, A and B, are too close to each other, thermostat A may cause thermostat B to shut off prematurely. Thus, the humidity can be high in one part of the premises and mold can start growing in the B duct system and around the air vents.

DRAIN PAN
Installing an air handler on the second floor or in the attic of a dwelling should be avoided. Leaks or overflow from the drain pan can cause major mold problems between floors.

Every month pour a cup of bleach into the drain line. This will prevent debris or sludge from accumulating and clogging the line causing water to back up inside your home. (WW)

WATER HEATER
Water heaters should never be put in attics. A one time or another they will leak. Set a WW in the drip pan.

CEILING FANS
Ceiling fans help reduce the humidity in the air. If you happen to live in a humid climate and have ceiling fans in your home, use them. If you don't have fans, think about installing a few. They help air circulate and they offer energy savings in both the cooling and heating seasons.

ATTIC

Twice a year check your attic for roof leaks and signs of rodents. Rodents in the attic may chew on ductwork. When leakage occurs, warm air from the attic and cool air from the A/C may cause condensation around the opening and mold will have a chance to grow inside the ductwork.

PINHOLE LEAKS

Older homes have copper pipes. Over time chemicals in water will corrode the pipes resulting in pinhole leaks. Some holes will seal themselves while others will get bigger causing water damage and mold.

Re-piping can be quite costly because it is labor intensive. The pipes are usually installed overhead. The process requires making holes in walls and ceilings to install the new CVPC pipes and it involves reconstruction of the drywall, re-texture and painting. Once completed the owner will have a brand new plumbing system, which should last a very long time.

Another alternative involves removing chemically the corrosion inside the existing pipes and then coating the inside of the pipes with an Epoxy based resin. The coating is supposed to seal the holes and prevent future pinholes. It is quite convenient. However, it is usually more costly than re-piping and the end result may be less than satisfactory because once completed you will still have an old plumbing system. It would behoove you to do your own research before choosing this alternative.

BASEMENT

Periodically check the sump pump to make sure it is in good working order and it discharges a good distance from the house. Having a battery backup system can be very useful in case power is lost. Place several WW units throughout the basement.

OUTSIDE YOUR HOME

CHECK FOR SMALL CRACKS
Make it a habit to walk around your home once a month to check your home for small cracks around windows and doors or in the stucco. Caulk any cracks immediately.

GUTTERS
Regularly clean the gutters of leaves and debris that may accumulate and cause water to infiltrate your home.

SPRINKER SYSTEM
Sprinkler heads should be two feet away from the house, at a minimum. Do not assume that they point away from the house. While the sprinklers are on, look to see if water hits the house. Sprinkler splash guards can be purchased at hardware stores.

PROPER DRAINAGE
The slope around the house should allow water to drain away from the house.

VINES
Do not put vines against outside walls. If you absolutely want the look, check with your hardware store for a trellis system that will allow for a gap between the vines and the structure. Vines keep moisture up against the house, and they destroy the outside stucco/paint or other coatings and will cause you trouble sooner rather than later.

PLANTER BOX
Do not install a planter box against the house unless it has a liner to allow water to drain to ground level and away from the house. Otherwise the humidity in the soil will migrate to the inside of the house and cause mold to grow.

TIMELY PREVENTIVE MAINTENANCE

INDOOR AIR QUALITY MONITORING
It is highly recommended that once a year random air samples be taken and analyzed to make sure that the air quality in relation to mold remains within the normal range.

ROOF INSPECTION
Have your roof inspected every year, and immediately after you experience a hurricane or major windstorm.

BEFORE LEAVING FOR VACATION
Do three things:
1. Leave your air conditioning on. See *MOLD PREVENTION: AIR CONDITIONING SYSTEM* – Northerners' Mistake
2. Turn off the power to the water heater,
3. Turn off the water at the main. If pipes break or leaks develop when you are away, your home won't be flooded. Obviously this will be bad for your lawn if you depend on city water and you are gone a while.

INFRARED SURVEYS – THERMAL IMAGING
Thermal imaging has become the avant-garde tool of predictive maintenance in commercial applications and is slowly gaining popularity in the private sector. By doing periodic audits of structures and systems, companies are becoming proactive so that potential problems can be averted. Now, private individuals are also taking advantage of this unique predictive safety maintenance tool to protect their own personal real estate. Infrared cameras are great tools in finding water intrusion in a home, building, or boat. Preventing water intrusion prevents mold.

There are two types of flat roofs – those that leak and those that do not leak yet. For small residential roofs an infrared survey can be performed (by walking) on the roof. For large commercial roofs an aerial infrared survey is best. It allows seeing the areas of moisture intrusion at a glance. It is also faster and more economical than on-roof surveys.

Benefits of aerial infrared surveys of commercial roofs:
- Accurate roof assessment for water intrusion
- Extension of roof life by periodic aerial surveys
- Buyer beware - pre-purchase moisture surveys
- Seller beware - pre-existing condition exclusion
- Forced warranty compliance by roofers
- Mold prevention through early water detection

Visual photograph of a roof

Infrared thermograph of a roof

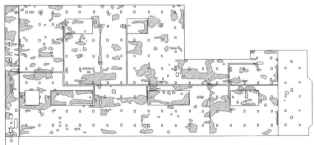

CADD drawing showing areas of water intrusion

Images courtesy of Stockton Infrared, Inc.

IMPROVING AIR QUALITY WITH PLANTS

INDOOR AIR POLLUTANTS

Research by Dr. Wolverton, one of the world's leading scientists on plants and indoor pollutants, has led to the recognition that plants can help clean the air of toxic chemicals in sick-building related illnesses. One of the most common toxins found in an indoor environment is Formaldehyde. It is not surprising that since this chemical is known to cause cancer in rodents, that it also causes many health problems in humans. The effects range from eye, nose and throat irritation to asthma, cancer, chronic respiratory diseases and neuropsychological problems.[20]

SOURCES OF CHEMICAL EMISSIONS

	FORMALDEHYDE	XYLENE / TOLUENE	BENZENE	TRICLOROETHYLENE	CLOROFORM	AMMONIA	ALCOHOLS	ACETONE
Adhesives	■	■	■				■	
Bioeffluents		■				■	■	
Blueprint machines						■		
Carpeting	■							
Caulking compounds	■						■	
Ceiling tiles	■						■	
Chlorinated tap water					■			
Cleaning products						■		
Computer VDU screens	■							
Cosmetics							■	■
Duplicating machines				■				
Electro-photographic printers		■						
Draperies	■							
Fabrics	■							
Facial tissues	■							
Floor coverings	■						■	
Gas stoves	■							
Grocery bags	■							
Microfiche developers						■		
Nail polish remover								■
Office correction fluid								■
Paints							■	
Paper towels	■							
Particle-board	■	■	■				■	
Permanent-press clothing	■							
Photo-copiers		■						
Plywood	■							
Pre-printed forms								■
Stains and varnishes	■	■	■				■	■
Tobacco smoke	■		■					
Upholstery	■							
Wall coverings		■					■	

Wolverton, B. C. *How To Grow Fresh Air,* Penguin, New York, 1996.

Off-gassing from building materials and new products will slowly become a thing of the past. In an effort to reduce indoor air pollution, the concept of "green building" has forced many manufacturers to produce environmental friendly materials that release low to zero volatile organic compounds (VOCs). The objective of the green building concept is to provide energy efficient buildings that are healthy for their occupants and good for the planet.

The push is on to develop "green" building products for home building and remodeling. To facilitate the exchange of ideas and as an educational opportunity, on a yearly basis, the National Association of Home Builders sponsors the *National Green Building Conference*. There, builders, remodelers, developers, architects, engineers, and other building professionals come together to share new ideas and products.

With a name like "green building" you would almost expect that it involves plants, but this is not the case. Builders construct buildings and gardeners grow greens. We can't expect builders to be concerned with plants nor can we expect gardeners to be concerned with buildings. It is thus up to the owners to embrace the work of Dr. Wolverton by bringing plants into homes and buildings, not only to help purify the air but also provide psychological well-being to the occupants.

It is said that indoor air pollution is one of the major threats to health. Considering that most people spend 90% of their time indoors, it would be an ideal solution to provide an indoor environment "that mimics the way that nature cleans the earth's atmosphere."[20]

There are hundreds of VOCs found in homes and buildings. Because formaldehyde is the predominant pollutant found indoors, Dr. Wolverton chose this particular toxin in his research as the standard for rating the ability of fifty plants to remove volatile organic compounds. The results of his study are found in the following table:

REMOVAL OF THE TOXIC GAS FORMALDEHYDE
BY HOUSEPLANTS IN µg PER HOUR

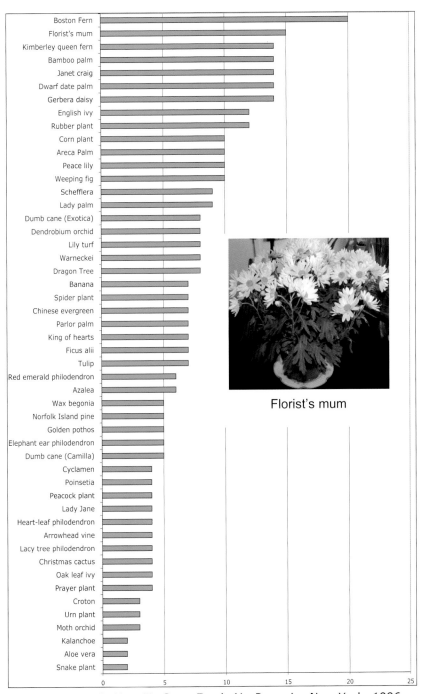

Florist's mum

Wolverton, B. C. *How To Grow Fresh Air,* Penguin, New York, 1996.

INVENTORY YOUR POSSESSIONS

INVENTORY YOUR POSSESSIONS

The Insurance Information Institute (iii) has made available a free program to help owners inventory personal possessions. This inventory is highly recommended for many reasons – when submitting a claim for damage suffered as a result of a natural disaster, a theft, or creating a will, to name a few. The software can be downloaded from its website: www.knowyourstuf.org

The Insurance Information Institute gives the following tips on taking a home inventory:

> This software will help you create a room-by-room inventory of your personal possessions. Having an up-to-date home inventory will help you:
> - Purchase enough insurance to replace the things you own;
> - Get your insurance claims settled faster;
> - Substantiate losses for your income tax return.

You can always make a list in a notebook and save receipts and photos in a file, but this software should make this task fun and simple. More importantly, with a click of a mouse, you can update your inventory as you buy or eliminate personal possessions.

GETTING STARTED

If you are just setting up a household, starting a home inventory can be relatively simple. You could even attach recent wedding registries to substantiate new possessions. But, if you have been living in your house for many years, this task may seem daunting. If you set aside an afternoon and get your entire household involved, it can be an enjoyable experience. In all cases, it is much easier to document your possessions before you suffer a loss from a fire, hurricane, burglary or other disaster, rather than having to document them in time of crisis.

BIG TICKET ITEMS

Make note of expensive items, such as jewelry, furs, and collectibles. Valuable items may need separate insurance. But, don't forget more commonplace items such as toys, CD's and clothing.

TAKING PHOTOGRAPHS

Along with the written information, consider adding photographs of your possessions, which can be done easily with a digital camera. You can also simply store your print photographs along with a copy of your inventory, or have the photographs scanned and the images saved to disk.

VIDEOTAPE YOUR INVENTORY

Walk through your house or apartment videotaping the contents. Remember to open drawers and closets. One advantage of videotaping your possessions is that you can narrate what you are filming.

PLAYING HIDE AND SEEK WITH MOLD

By

Rachelle Dobbs and Greg Stockton

Excerpt from a paper presented at the IRINFO, Infraspection Institute, January 2005, Orlando Florida on the use of infrared thermography in mold detection.

Abstract

Four hurricanes hit the State of Florida in six weeks in 2004: Charley, Frances, Ivan and Jeanne. Charley struck on Friday, August 13th, Frances followed three weeks later, Ivan came ashore on September 16th and Jeanne on Sept. 26th. Huge amounts of rain, coupled with wind gusts over 100 miles per hour caused tremendous damage. More than one in five Florida homes suffered some sort of damage, according to news reports. Insurance claims topped two million. State officials estimate insured losses at $18 billion.

The damage from these hurricanes provided very good conditions for water intrusion and in some cases, mold growth. Infrared (IR) thermographers and mold inspectors can be effective partners in quantifying damage and detecting mold, as long as both understand the advantages and limitations of their respective fields. The expertise of both is enhanced with microbial sampling, which scientifically establishes the presence or absence of mold. Understanding construction, water intrusion and mold behavior all help selecting the best location for obtaining accurate mold levels, remediating the mold and in repairing the damage.

Introduction

Interestingly, new homes built since 2000 seemed to have had the most water intrusion and moisture problems in both frequency and severity. A 2003 survey conducted by The Orlando Sentinel newspaper and WESH-TV found that 60 percent of homes built in the Orlando area in 2001 had cracks around windows and doors, in the foundations and in the stucco or stucco-like finishes on the block walls. 1

Two main types of water intrusion occurred during the hurricanes. One came from the top, and one from the side. Leaks coming from the top came from compromised roofs when shingles or parts of the roof were blown off and relentless rain poured into homes. Water intrusion from the sides, soon termed "water seepage" came mostly from concrete block walls (especially in the direct path of the hurricanes) and from around windows that were not properly sealed or flashed. Most water intrusion problems were caused by either improperly painted/sealed exterior walls, by cracks in the stucco-like coating applied on the walls or by sloppy mortar joints between the blocks. [Seepages through improperly sealed block walls happen naturally because of the capillary action of the concrete block itself.] The combination of large amounts of rain and high winds pounding at a 90° angle, resulted in rainwater pouring into homes by the gallon; flooding living rooms, bedrooms and kitchens, warping walls, cabinets, wood floors and soaking carpets.

Within a few days some victims of the 2004 Florida hurricanes noticed a musty smell. Within weeks, mold inspectors and mold remediators were inundated with phone calls. Often, the owners of new homes called their builders to complain about their "leaky houses" to the point that many the builders refused to answer their phones. A few homeowners took it upon themselves to obtain proof of water intrusion by hiring infrared thermographers to document the location of the moisture inside the building structure. With irrefutable proof in hand, they felt that they could convince their builders to correct building defects. But generally, the builders brushed off their complaints, claiming an "Act of God". Homeowners then turned to their insurance companies, which categorically turned down the claims on the basis that construction defects were not covered. When they turned to city and county governments – which had issued building permits, they found no help since there were no provisions in the building code for waterproofing. Since then, some cities are instituting an ordinance to force homebuilders to waterproof homes. 2

Water Intrusion and Moisture
We must differentiate between water intrusion and moisture in relation to mold and water damage. Water intrusion implies that the building has been compromised and water is coming from somewhere, either from the roof, the walls, the basement, or from a mechanical or plumbing leak. Moisture inside the structure is a result of any one or a combination of factors, i.e., a water intrusion problem, a thermal envelope problem, a ventilation problem and/or an HVAC system problem. Swift action of the homeowner or building owner can make a huge difference after an event, like a hurricane. Drying out the building should be done immediately because mold problems only compound with time. If mold has already settled in, using powerful fans to dry structures will only spread mold spores and cross-contaminate other areas of the building. Water vapor can increase moisture content in the building materials, thus encouraging mold growth. The faster the infrared thermographer performs the survey after the event, the more accurate the assessment will be in relation to water damage. Same with the mold inspection. The faster the mold inspectors perform their assessment, the less severe the mold contamination will be.

Infrared Thermography Cannot Be Used To Detect Mold
IR thermography can be used to find moisture in building materials, see thermal and moisture envelope problems that can create the right conditions for mold growth in buildings, and to see if the active HVAC system is creating problems that can contribute to mold growth in buildings. In extreme cases where the building materials are deteriorated to the point where the mass of the material is affected, IR can be used; however a very sensitive IR imager and advanced IR techniques are usually required. Infrared thermographic surveys work well to find moisture in building materials when there is good thermal contrast due to the evaporative

cooling effect and when temperature differentials are at their peak, but diminish as soon as the materials dry. **Read these words and heed them: <u>Infrared thermography cannot be used to detect mold</u>.** Mold does not exhibit an exothermic reaction that can be seen with an infrared camera by walking around a building (See Figures below).

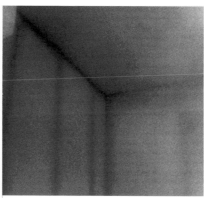

| Visual image with no visible signs of mold. | Infrared image with no moisture indicated. |

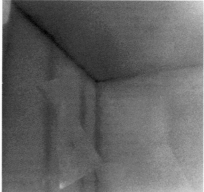

Visual image with mold present.

Infrared image with wallpaper removed
There is still no signs of moisture.

Clear thermographic images of water intrusion and moisture problems were made possible shortly after the hurricanes, due to the moisture still being present on building surfaces. One big advantage is that large and inaccessible areas can be surveyed quickly and efficiently. Moisture problems in ceilings, which are not readily accessible with hand-held moisture meters, can be readily seen even with no stains present. Large areas of walls, windows and doors can be well-documented with infrared images. In a report, visual and infrared images are much more convincing than, say, "the moisture meter registered 'red' around the window". The old adage "a picture is worth a thousand words" is true.

Infrared thermographers helped detect moisture in building materials and provided clients with a visual record of anomalies consistent with moisture intrusion, but as the months went by and building materials dried out, thermographic surveys were less useful and more difficult to perform. Infrared surveys are not effective when materials that were previously wet are dry, since there is no longer any temperature differential to detect. Some thermographers had to resort to 'flood testing' methods [wetting the walls with a water hose on the outside, while taking thermal images inside]. One might believe that by locating moisture, mold can be located. This is not always the case. Here are a few scenarios that occurred shortly after the hurricanes:

Scenario 1
Water/moisture came from the roof onto the ceiling and ran down the inside of the walls and came out onto tile flooring. An infrared survey showed hardly any moisture on the ceiling. The homeowners had sponged up the water/moisture right away and used fans to dry out the wall. The building materials dried quickly before mold had a chance to grow.

Scenario 2
Same conditions as above, except that the homeowners had carpeting and had not used a fan. Water/moisture in the carpet kept the environment damp for a long time and as a result the drywall wicked up water/moisture from the carpet and stayed moist long enough to allow mold to grow.

Mold then grew on the back of the drywall and on the baseboard. Mold was also apparent from inside the room.

Scenario 3
Water/moisture came from the roof, ran along the side of a cathedral ceiling then down the inside of the wall. An infrared survey showed that the ceiling was wet as well as the bottom of the wall, but IR could not see the behind the walls, where the mold was growing. Also, the carpet kept the drywall moist and mold was found growing about two feet up from the bottom of the drywall.

Scenario 4
Water/moisture came from the roof onto the ceiling only and for some reason the water/moisture was able to pool. A water/moisture stain could be seen. An infrared survey showed that the ceiling was wet. Damp insulation kept the environment wet and mold started to grow on the ceiling and the blown-in insulation.

Mold detection is complex because mold can hide in the most inconspicuous of places and its behavior is sometimes unpredictable. Infrared thermography is a great tool to detect wet media. However, it has limitations in mold detection, mainly:
- If the medium was once wet and is now dry, no anomalies will be seen even though mold might have grown on the other side.

- If the wet area dried within 48 hours, the chances for mold to grow will be minimal.

- A wet spot may be only a wet spot. Water/moisture, taking the path of least resistance, may end up far away from the original source and mold may therefore grow far away from the point of the original moisture intrusion or on a surface.

Conclusions
Infrared thermographers must recognize the limitation of infrared surveys in mold detection. It is always best to

recommend that the client also hire an expert mold inspector. In all types of buildings, moisture information gathered by an infrared thermographer is extremely valuable to the mold inspector because it narrows down the search for mold, so long as the IR survey is performed quickly after the event. This saves the client time and money. Experts in both fields are needed – one dealing in moisture detection and the other in mold detection. The final objective is to help the client determine whether and where water damage has occurred and if so, whether there is a mold problem in the building. The cost of hiring both experts is minimal when the owner considers the preservation of structural integrity of the building and safeguarding the health of its occupants.

References

1. Tracy, Dan. "City Looks To Tighten Rules for Builders" The Orlando Sentinel, 11 Nov. 2004, final edition: sec. A: 1, 8

2. Tracy, Dan, et al. "Home Builders Are Told To Add Waterproofing" The Orlando Sentinel, 12 Nov. 2004, final edition: sec. B: 3

A copy of the complete paper can be found at:
www.MoldDetectionExperts.com/articles.html

FUNGAL LIBRARY

Here we have included the types of mold most commonly found in homes.

Information and pictures are reprinted with permission from The Environmental Microbiology Laboratory, Inc. (EMlab™)

More information on other types of mold can be obtained from their extensive mold library at:
www.emlab.com/app/fungi/Fungi.po

MOLD FOUND IN ABUNDANCE OUTDOORS

The following mold genera are among the most common mold spores found outdoors:

- *Penicillium*
- *Aspergillus*
- *Cladosporium*
- *Basidiospores*

In spite of their normal occurrence we expect to find them indoors in lesser or no greater amounts than outdoors. The levels of mold spores vary upon the geographical location, daily weather conditions, the vegetation (plants and tress) found nearby, inside/outside air exchange rates, and other factors.

MOLDS SELDOM FOUND OUTDOORS

The following mold types are seldom found outdoors:
- *Stachybotrys*
- *Chaetomium*

These are generally found in very low numbers outdoors. Consequently their presence indoors, even in relatively low numbers, is often an indication that these molds are originating from growth indoors and are often the clearest indicator of a mold problem. When water availability is high for prolonged periods on organic materials, *Stachybotrys* may gradually become the predominant mold.

PENICILLIUM/ASPERGILLUS

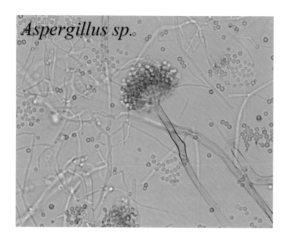
Aspergillus sp.

Penicillium and Aspergillus are two separate genera of molds whose spores are visually similar and are grouped together in some laboratory analyses.

MODES OF DISSEMINATION:
Penicillium/Aspergillus produce dry spore types that are easily dispersed through the air by wind. These fungi serve as a food source for mites, and therefore can be dispersed by mites and various insects as well.

WHERE IT IS FOUND OUTDOORS:
Penicillium/Aspergillus are found in soils, decaying plant debris, compost piles, fruit rot and some petroleum-based fuels.

WHERE IT IS FOUND INDOORS:
Penicillium/Aspergillus are found throughout the home. They are common in house dust, growing on wallpaper, wallpaper glue, decaying fabrics, wallboard, moist chipboards, and behind paint. They have also been isolated from blue rot in apples, dried foodstuffs, cheeses, fresh herbs, spices, dry cereals, nuts, onions, and oranges.

Potential toxin production.

CLADOSPORIUM

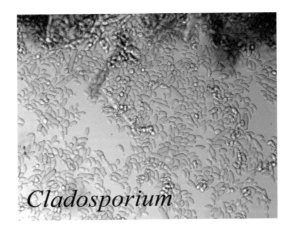

Cladosporium

MODES OF DISSEMINATION:

Cladosporium produces dry spores that are formed in branching chains. Spores are released by the agitation of the spore-bearing hyphae as they dry. Thus, the spores are most abundant in dry weather.

WHERE IT IS FOUND OUTDOORS:

Cladosporium is found in a wide variety of soils, in plant litter, and on old and decaying plants and leaves. Some species are plant pathogens

WHERE IT IS FOUND INDOORS:

Cladosporium can be found anywhere indoors, including textiles, bathroom tiles, wood, moist windowsills, and any wet areas in a home. Some species of *Cladosporium* grow at temperatures near or below 0(C) / 32(F) and can often be found on refrigerated foodstuffs and even frozen meat.

Potential toxin production.

BASIDIOSPORES

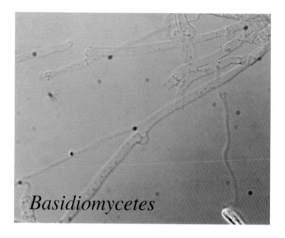

Basidiomycetes

These particular fungi are related to mushrooms. Their spores are abundant outdoors following a rainy day.

MODES OF DISSEMINATION:

Many types of basidiospores are actively released into the air during periods of high humidity or rain. Once the spores are expelled into the air, they are easily dispersed by wind.

WHERE THEY ARE FOUND OUTDOORS:

Basidiomycetes are very common outdoors and can be found in gardens, forests, grasslands, and anywhere there is a substantial amount of dead organic material. They are also found on or near plants and some are known to be plant pathogens.

WHERE THEY ARE FOUND INDOORS:

Basidiospores found indoors typically come from outdoor sources and are carried inside by airflow or on clothing. Certain kinds of *basidiomycetes* can grow indoors, such as those that cause "dry rot", which can cause structural damage to wood. Occasionally, other *basidiomycetes* such as mushrooms can be found indoors, but this is not common. Generally, *basiodiomycetes* require wet conditions for prolonged periods in order to grow indoors.

STACHYBOTRYS

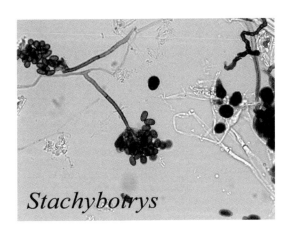

Stachybotrys

MODES OF DISSEMINATION:
Stachybotrys produces wet slimy spores and is commonly dispersed through water flow, droplets, or insect transport, less commonly through the air. *Stachybotrys* mold spores do not aerosolize easily.

WHERE IT IS FOUND OUTDOORS:
Stachybotrys is found in soils, decaying plant debris, decomposing cellulose, leaf litter and seeds.

WHERE IT IS FOUND INDOORS:
Stachybotrys is common indoors on wet materials containing cellulose such as wallboard, jute, wicker, straw baskets, and other paper materials.

Potential toxin production.

CHAETOMIUM

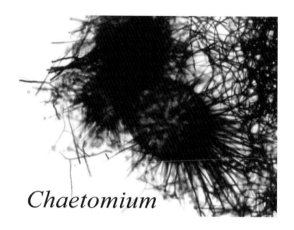

Chaetomium

MODES OF DISSEMINATION:
Chaetomium spores are formed inside fruiting bodies. The spores are released by being forced out through a small opening in the fruiting body. The spores are then dispersed by wind, water drops, or insects.

WHERE IT IS FOUND OUTDOORS:
Chaetomium can be found in soil, on various seeds, cellulose substrates, dung, woody materials and straw.

WHERE IT IS FOUND INDOORS:
Chaetomium can grow in a variety of areas indoors, but is usually found on cellulose-based or woody materials in the home. It is very common on sheetrock paper that is or has been wet.

Potential toxin production.

MOLD REMEDIATION IN BIBICAL TIMES

Cleansing From Mildew
Leviticus, Chapter 14:33-53

33 The LORD said to Moses and Aaron, 34 "When you enter the land of Canaan, which I am giving you as your possession, and I put a spreading mildew in a house in that land, 35 the owner of the house must go and tell the priest, 'I have seen something that looks like mildew in my house.

36 The priest is to order the house to be emptied before he goes in to examine the mildew, so that nothing in the house will be pronounced unclean. After this the priest is to go in and inspect the house. 37 He is to examine the mildew on the walls, and if it has greenish or reddish depressions that appear to be deeper than the surface of the wall, 38 the priest shall go out the doorway of the house and close it up for seven days.

39 On the seventh day the priest shall return to inspect the house. If the mildew has spread on the walls, 40 he is to order that the contaminated stones be torn out and thrown into an unclean place outside the town. 41 He must have all the inside walls of the house scraped and the material that is scraped off dumped into an unclean place outside the town. 42 Then they are to take other stones to replace these and take new clay and plaster the house.

43 "If the mildew reappears in the house after the stones have been torn out and the house scraped and plastered, 44 the priest is to go and examine it and, if the mildew has spread in the house, it is a destructive mildew; the house is unclean. 45 It must be torn down-its stones, timbers and all the plaster-and taken out of the town to an unclean place. 46 "Anyone who goes into the house while it is closed up will be unclean till evening. 47 Anyone who sleeps or eats in the house must wash his clothes.

48 "But if the priest comes to examine it and the mildew has not spread after the house has been plastered, he shall pronounce the house clean, because the mildew is gone. 49 To purify the house he is to take two birds and some cedar wood, scarlet yarn and hyssop. 50 He shall kill one of the birds over fresh water in a clay pot. 51 Then he is to take the cedar wood, the hyssop, the scarlet yarn and the live bird, dip them into the blood of the dead bird and the fresh water, and sprinkle the house seven times. 52 He shall purify the house with the bird's blood, the fresh water, the live bird, the cedar wood, the hyssop and the scarlet yarn. 53 Then he is to release the live bird in the open fields outside the town. In this way he will make atonement for the house, and it will be clean."

REFERENCES

1. Cook, Gareth " Astronauts Vs. Fungus – Orbiting Spacecraft Turns Out To Be Food For Aggressive Mold." The Boston Globe, 1 Oct. 2000; Sec: A01.

2. "Terracotta Army Battles New Enemy" BBC. England, 18 Sep. 2000.

3. Harriman, Brundrett & Kittler. Humidity Control Design Guide. 102, ASHRAE, American Society of Heating, Refrigerating and Air Conditioning Engineers, Atlanta, GA, December 2001.

4. Burge, Harriet. Bioaerosols. 103. Lewis Publishers, Boca Raton, FL, 1995.

5. Burge, Harriet. Bioaerosols. 107. Lewis Publishers, Boca Raton, FL, 1995.

6. Dumanov Joseph. "Should Dogs Be Used To Inspect For Toxic Mold?" Dec. 2005.
   <http://www.njmoldinspection.com/vetmycology/vetasper.html>

7. Mann, Arnold. "MOLD: A Health Alert" USA Weekend Magazine. 5 Dec. 1999; 8-9.

8. "St. Anthony's Fire - Ergotism". Nov. 2005
   <www.medicinenet.com/script/main/art.asp?articlekey=14891>

9. Kendrick, Bryce. The Fifth Kingdom. 216, Focus, Newburyport, MA, 2000.

10. Ponikau, J., Sherris, D., Kern, E., Homburger, H., Frigas, E., Gaffey, T., Roberts, G. The Diagnosis and Incidence of Allergic Fungal Sinusitis. Mayo Clinic Proc 1999; 74:877-884.

11. Lethbridge-Cejku M, Schiller JS, Bernadel L. Summary health statistics for U.S. adults; National Health Interview Survey, 2002. National Center for Health Statistics. Vital Health Stat. 2004; 10:23.

12. Sasama, J., Sherris, D., Shin, S, Kephart, G., Kern, E., Ponikau, J. <u>New Paradigm for the Roles of Fungi and Eosinophils in Chronic Rhinosinusitis.</u> Current Opinion in Otolaryngology & Head and Neck Surgery 2005, 13:2-8.

13. TheIEQReview, February 2, 2005, Volume 1, Issue 145. "Surgeon General's Office Issues Conflicting Message on "Toxic Mold." by PRWEB, January 27, 2005.

14. Storey, E and Als. <u>Guidance for Clinicians on the Recognition and Management of Health Effects Related to Mold Exposure and Moisture Indoors.</u> University of Connecticut Health Center, Sep. 30, 2004, 23-24.

15. Lstiburek, Joseph. <u>Moisture Control Handbook, New Low-rise, Residential Construction</u>. US Department of Energy, Oct. 1991, xiv.

16. Wolverton, B. C., Johnson, A., Bounds, K., <u>Interior Landscape Plants for Indoor Air Pollution Abatement</u>. National Aeronautics and Space Administration, John C. Stennis Space Center, Science and Technology Laboratory, Stennis Space Center, MS, September 15, 1989.

17. "Indoor Air Facts No. 4 (revised): Sick Building Syndrome (SBS)." April 1991 <www.epa.gov/iaq/pubs/sbs.html>

18. "Culturable vs. non-culturable methods." Sep. 2005 <u><www.emlab.com/s/sampling/Sampling.html></u>

19. Dillon, H. Heinsohn, P. and Miller, J.(eds) <u>Field Guide for the Determination of Biological Contaminants in Environmental Samples</u>, Biosafety Committee, American Industrial Hygiene Association, Fairfax, VA, 1996.

20. Wolverton, B. C. <u>How To Grow Fresh Air</u>. 8-23. Penguin Books, New York, NY, 1996.

OTHER SOURCES:

American Academy of Allergy Asthma & Immunology
www.aaaai.org

American Industrial Hygiene Association
www.aiha.org

American Institute For Conservation of Historic & Artistic Works
http://aic.stanford.edu/

American Lung Association
www.lungusa.org

American Society of Heating, Refrigerating and Air Conditioning Engineers, Inc.
www.ashrae.org

Architecture and Building Science
www.buildingscience.com

Center Watch – Clinical Trials: Asthmas (Pediatric)
www.centerwatch.com/patient/studies/cat442.html

Children Environmental Health – Resource Guide
www.cehn.org/cehn/resourceguide/ala.html

Department of Environmental Health & Safety (DEHS) at
the University of Minnesota
www.dehs.umn.edu/iaq

Disaster Help (DHELP)
www.disasterhelp.gov

Empowering the Policy Holder (POA)
www.policyholdersofamerica.org

Environmental Microbiology Laboratory – Fungal Library
www.emlab.com/app/fungi/fungi.po

Federal Emergency Management Agency
www.fema.gov
www.fema.gov/hazards/floods/whatshouldidoafter.shtm

Free publications
http://oaspub.epa.gov/webi/meta_first_new2.try_these_first

Home inventory software (free)
www.knowyourstuf.org

Hurricane Insurance Information Center
www.disasterinformation.org

Institute for Environmental Assessment (IEA), Minneapolis,
Minnesota
www.ieainstitute.com

Mold in Schools
www.edfacilities.org/rl/Mold.cfm

National Aeronautics and Space Administration
www.nasa.gov

National Association of Home Builders
www.nahb.org

National Institute of Allergy and Infectious Diseases
www.niaid.nih.gov

National Institute of Standards and Technology
www.nist.gov/public_affairs/guide/index.htm

National Library of Medicine's Search Service
www.ncbi.nlm.nih.gov

New York City Department of Health and Mental Hygiene -
Guidelines on Assessment and Remediation of Fungi in
Indoor Environments
www.nyc.gov/html/doh/html/epi/moldrpt1.shtml

University of Minnesota Environmental Health and Safety
www.dehs.umn.edu/iaq/

U.S. Department of Housing and Urban Development
www.hud.gov

U.S. Department of Labor Occupational Safety & Health
Administration
www.oshaslc.gov/SLTC/respiratoryprotection/index.html

U.S. Environmental Protection Agency
www.epa.gov

Wolverton Environmental Services. Plants: the solution to
indoor air pollution
www.wolvertonenvironmental.com

World Allergy Organization
www.worldallergy.org